ASSESSMENT OF RADIOACTIVE CONTAMINATION, EXPOSURES AND COUNTERMEASURES IN URBAN ENVIRONMENTS

The following States are Members of the International Atomic Energy Agency:

AFGHANISTAN
ALBANIA
ALGERIA
ANGOLA
ANTIGUA AND BARBUDA
ARGENTINA
ARMENIA
AUSTRALIA
AUSTRIA
AZERBAIJAN
BAHAMAS, THE
BAHRAIN
BANGLADESH
BARBADOS
BELARUS
BELGIUM
BELIZE
BENIN
BOLIVIA, PLURINATIONAL STATE OF
BOSNIA AND HERZEGOVINA
BOTSWANA
BRAZIL
BRUNEI DARUSSALAM
BULGARIA
BURKINA FASO
BURUNDI
CABO VERDE
CAMBODIA
CAMEROON
CANADA
CENTRAL AFRICAN REPUBLIC
CHAD
CHILE
CHINA
COLOMBIA
COMOROS
CONGO
COOK ISLANDS
COSTA RICA
CÔTE D'IVOIRE
CROATIA
CUBA
CYPRUS
CZECH REPUBLIC
DEMOCRATIC REPUBLIC OF THE CONGO
DENMARK
DJIBOUTI
DOMINICA
DOMINICAN REPUBLIC
ECUADOR
EGYPT
EL SALVADOR
ERITREA
ESTONIA
ESWATINI
ETHIOPIA
FIJI
FINLAND
FRANCE
GABON
GAMBIA, THE
GEORGIA
GERMANY
GHANA
GREECE
GRENADA
GUATEMALA
GUINEA
GUYANA
HAITI
HOLY SEE
HONDURAS
HUNGARY
ICELAND
INDIA
INDONESIA
IRAN, ISLAMIC REPUBLIC OF
IRAQ
IRELAND
ISRAEL
ITALY
JAMAICA
JAPAN
JORDAN
KAZAKHSTAN
KENYA
KOREA, REPUBLIC OF
KUWAIT
KYRGYZSTAN
LAO PEOPLE'S DEMOCRATIC REPUBLIC
LATVIA
LEBANON
LESOTHO
LIBERIA
LIBYA
LIECHTENSTEIN
LITHUANIA
LUXEMBOURG
MADAGASCAR
MALAWI
MALAYSIA
MALI
MALTA
MARSHALL ISLANDS
MAURITANIA
MAURITIUS
MEXICO
MONACO
MONGOLIA
MONTENEGRO
MOROCCO
MOZAMBIQUE
MYANMAR
NAMIBIA
NEPAL
NETHERLANDS, KINGDOM OF THE
NEW ZEALAND
NICARAGUA
NIGER
NIGERIA
NORTH MACEDONIA
NORWAY
OMAN
PAKISTAN
PALAU
PANAMA
PAPUA NEW GUINEA
PARAGUAY
PERU
PHILIPPINES
POLAND
PORTUGAL
QATAR
REPUBLIC OF MOLDOVA
ROMANIA
RUSSIAN FEDERATION
RWANDA
SAINT KITTS AND NEVIS
SAINT LUCIA
SAINT VINCENT AND THE GRENADINES
SAMOA
SAN MARINO
SAUDI ARABIA
SENEGAL
SERBIA
SEYCHELLES
SIERRA LEONE
SINGAPORE
SLOVAKIA
SLOVENIA
SOMALIA
SOUTH AFRICA
SPAIN
SRI LANKA
SUDAN
SWEDEN
SWITZERLAND
SYRIAN ARAB REPUBLIC
TAJIKISTAN
THAILAND
TOGO
TONGA
TRINIDAD AND TOBAGO
TUNISIA
TÜRKİYE
TURKMENISTAN
UGANDA
UKRAINE
UNITED ARAB EMIRATES
UNITED KINGDOM OF GREAT BRITAIN AND NORTHERN IRELAND
UNITED REPUBLIC OF TANZANIA
UNITED STATES OF AMERICA
URUGUAY
UZBEKISTAN
VANUATU
VENEZUELA, BOLIVARIAN REPUBLIC OF
VIET NAM
YEMEN
ZAMBIA
ZIMBABWE

The Agency's Statute was approved on 23 October 1956 by the Conference on the Statute of the IAEA held at United Nations Headquarters, New York; it entered into force on 29 July 1957. The Headquarters of the Agency are situated in Vienna. Its principal objective is "to accelerate and enlarge the contribution of atomic energy to peace, health and prosperity throughout the world".

IAEA-TECDOC-2096

ASSESSMENT OF RADIOACTIVE CONTAMINATION, EXPOSURES AND COUNTERMEASURES IN URBAN ENVIRONMENTS

REPORT OF WORKING GROUP 2

MODELLING AND DATA FOR RADIOLOGICAL IMPACT ASSESSMENTS (MODARIA II) PROGRAMME

INTERNATIONAL ATOMIC ENERGY AGENCY
VIENNA, 2025

For further information on this publication, please contact:

Waste and Environmental Safety Section
International Atomic Energy Agency
Vienna International Centre
PO Box 100
1400 Vienna, Austria
Email: Official.Mail@iaea.org

Printed by the IAEA in Austria
July 2025
https://doi.org/10.61092/iaea.6tch-1gpb

IAEA Library Cataloguing in Publication Data

Names: International Atomic Energy Agency.
Title: Assessment of radioactive contamination, exposures and countermeasures in urban environments / International Atomic Energy Agency.
Description: Vienna : International Atomic Energy Agency, 2025. | Series: IAEA TECDOC series, ISSN 1011-4289 ; no. 2096 | Includes bibliographical references.
Identifiers: IAEAL 25-01770 | ISBN 978-92-0-114825-4 (paperback : alk. paper) | ISBN 978-92-0-114725-7 (pdf)
Subjects: LCSH: Radioactive pollution. | Radioactive pollution — Analysis. | Radioactive Pollution — Management. | Radioactive pollution — Environmental aspects.

FOREWORD

The IAEA has been organizing international programmes for testing models for the transfer of radionuclides in the environment and the estimation of radiation exposures since the 1980s. These programmes have contributed to a general improvement in such models, including improvements in the associated data and advancements in the capabilities of modellers in Member States. The IAEA publications on this subject over the past several decades demonstrate the comprehensive nature of these programmes and document the associated advances that have been made.

In 2012 the IAEA launched a programme entitled Modelling and Data for Radiological Impact Assessments (MODARIA). The original programme (MODARIA I) ran until 2015. From 2016 to 2019 the IAEA organized a follow-up programme, MODARIA II, where seven working groups continued much of the work of MODARIA I. This publication describes the activities carried out during MODARIA II by Working Group 2, which continued the work on the assessment of exposures and countermeasures in urban environments that began in MODARIA I. The related MODARIA I activities carried out previously are described in IAEA-TECDOC-2001, published in 2022.

The IAEA is grateful to all those who participated in Working Group 2 of the MODARIA II programme, in particular K. Thiessen (United States of America) as the working group leader. The IAEA officers responsible for this publication were J. Brown and T. Yankovich of the Division of Radiation, Transport and Waste Safety.

CONTENTS

SUMMARY

In recent years there has been a development in the complexity of models and approaches to effectively assess the dispersion of radionuclides in an urban environment and the associated radiation exposures of people following an atmospheric release. Through the Environmental Modelling for Radiation Safety (EMRAS and EMRAS II) and Modelling and Data for Radiological Impact Assessments (MODARIA I and MODARIA II) programmes, the IAEA has facilitated knowledge sharing on this topic, including through model intercomparison, testing and development. This publication describes the work undertaken by Working Group 2, Assessment of Exposures and Countermeasures in Urban Environments of the IAEA's Modelling and Data for Radiological Impact Assessments (MODARIA II) programme (2016–2019). The work carried out was a natural continuation of that completed during the first phase of the IAEA's MODARIA programme (2012–2015) by Working Group 2 on the same theme. In MODARIA II, the degree of complexity of the models used and the model intercomparison and testing was increased to address more complex dispersion and public exposure situations.

The objective of WG2 was to test and improve the capabilities of models used in assessment of radioactive contamination in urban settings, including dispersion and deposition events, short term and long term contaminant redistribution following deposition events, and potential countermeasures or remediation efforts for reducing human exposures.

Working Group 2 undertook six modelling exercises across three major areas of activity during MODARIA II:

(1) Two modelling exercises applicable to contaminant transport inside an urban area (short range);
(2) Three modelling exercises applicable to contaminant transport to urban areas from an external location (mid-range);
(3) A modelling exercise to assess the distribution of external doses to members of the public following the deposition of radionuclides in an urban environment.

The first modelling exercise was a short range atmospheric dispersion exercise similar to two short range modelling exercises carried out during the EMRAS II and MODARIA I programmes. This exercise was based on data from a field test performed by the Czech National Radiation Protection Institute on a test area at the Boletice military training area, in the Czech Republic. The exercise was designed to enable comparison of model predictions with measurements of surface contamination up to 200 m downwind.

In this field test, a short lived radionuclide (^{140}La) in liquid form was spread by detonation of a small amount of explosive in an open field (flat terrain) in an omnidirectional explosion. Measurements included dose rates, surface contamination, and activity concentrations in air. Participants in the modelling exercise were asked to submit predictions for surface contamination (Bq/m^2) at specified locations (defined by a coordinate system).

Model predictions agreed with the measurements in terms of the general direction of the plume and the resulting deposition, although all models predicted the maximum deposited activity at a greater distance from the dispersion point than the location of the maximum measured deposition. Visual comparison of contour plots of measured and predicted deposition provided a useful way to compare the predictions with the measurements.

The second short range exercise was undertaken by two participants in order to follow up on a previous exercise conducted during the MODARIA I programme. The goal of both exercises was a comparison of the participants' respective decision support models using a hypothetical release located in a dense urban area in Munich, Germany. The exercise demonstrated the effects of different building arrangements on activity concentrations in air and deposition in the vicinity of the buildings and in downwind areas, for a set of relatively simple conditions and two types of release (explosion and continuous release). The predicted activity concentrations in air and the resulting deposition were slightly higher for the explosion, for which the effective release point was higher. The best agreement between the two models appeared to correspond to the scenario for which the building effects appeared to have the least impact.

Three mid-range atmospheric dispersion exercises were carried out during the MODARIA II programme; these exercises are intended to be applicable to situations such as nuclear power plant accidents, in which contamination from an accident site could be transported to urban areas some distance away. One of these exercises was based on the Šoštanj Thermal Power Plant in Slovenia, adding a third case to the two that were modelled during the MODARIA I programme. This case was selected to provide a meteorological situation of intermediate complexity between the two cases previously used. The contaminant was emitted from two nearby sources, spreading first towards the west or northwest and later towards the east or northeast. Two participants provided model predictions for this exercise. Both models predicted plumes generally toward the northwest and northeast, but differed in the predicted contaminant concentrations. Peak concentrations at specific locations (monitoring stations) were difficult to model, but allowing for spatial and temporal error in one model improved the predictions obtained.

The second mid-range exercise was based on one of the two cases from the Šoštanj Thermal Power Plant that were used during the MODARIA I programme. However, this exercise compared a diagnostic approach based on real time meteorological measurements with a prognostic approach based on predicted meteorological conditions obtained from forecasting models. Two participants provided results of prognostic simulations of meteorological variables, using two versions of the same weather model and two sets of input and boundary conditions. The most important weather variables with respect to modelling of atmospheric dispersion include air temperature, wind speed, global solar radiation, and precipitation. Forecasts of wind speed were generally too high (overpredictions), while air temperatures were slightly underpredicted.

The third mid-range exercise was based on a set of monitoring data for ^{41}Ar near the NRU (National Research Universal) research reactor at Chalk River, in Canada. Participants were provided with a site description, source term data, and meteorological data and were asked to predict the gamma dose rate at a downwind monitoring station for comparison with measurements. Five models were used in the exercise, including one Gaussian plume model and four Lagrangian particle models. Use of forecast meteorology versus measured meteorology gave the largest differences between the model simulations. The differences between measured and predicted gamma dose rates may be explained by the distance between the meteorological station and either the emissions source or receptor location, or by the lack of time dependent information about the source term. Comparison of time integrated results helped in addressing uncertainties in timing of the plume, and comparison of contour plots helped in addressing uncertainties in prediction of an endpoint at a specific location.

The final modelling exercise focused specifically on urban contamination in Fukushima City in Japan and doses to people from external radiation following the Fukushima Daiichi Nuclear Power Plant accident in 2011. Input data included surface contamination levels, information on building construction, and behavioural data. Participants were asked to predict the distribution of annual effective dose from external exposure for two groups of persons (defined in terms of occupation, i.e. indoor workers and outdoor workers) and doses to representative individual persons (calculated by a deterministic approach and a probabilistic approach). Five participants provided results for this exercise, using two main types of modelling approach (location-based and element-based). Both types of model were able to reproduce the distributions of measured doses. In a few cases, the predicted 95th percentiles of the dose distributions were less than the 95th percentiles of the measured doses (the 95th percentile is used by the ICRP in its concept of the representative person. So, in this exercise, the uncertainty and variability included within the assessments did not fully account for the actual situation. Deterministic results of external doses were always higher than the probabilistic results obtained with the same model.

For each of the modelling exercises, comparisons have been made between model predictions, and (where applicable) between model predictions and measurements. This enables the differences in model capabilities, the focus of different modelling groups, different types of dispersion model and processes, and the interpretation of input information, assumptions, and selection of parameter values to be evaluated. To understand both the similarities and differences in results, it is necessary to understand all these factors that can influence the model results. Comparison and discussion of predictions from several models provides a valuable opportunity to better understand the model results and to reduce errors in the modelling. The range of results for each of the exercises gives an idea of the overall level of uncertainty that can be expected for a given type of situation being modelled.

1. INTRODUCTION

1.1. BACKGROUND OF THE MODARIA II PROGRAMME

The IAEA organized a programme from 2016 to 2019, entitled Modelling and Data for Radiological Impact Assessments (MODARIA II), which had the general aim of enhancing the capabilities of Member States to simulate radionuclide transfer in the environment and, thereby, to assess exposure levels of the public and in the environment in order to ensure an appropriate level of protection from the effects of ionizing radiation associated with radionuclide releases and from existing radionuclides in the environment.

The following topics were addressed in seven working groups:

- Working Group 1: Assessment and Decision Making of Existing Exposure Situations for NORM and Nuclear Legacy Sites
- Working Group 2: Assessment of Exposures and Countermeasures in Urban Environments
- Working Group 3: Assessments and Control of Exposures to the Public and Biota for Planned Releases to the Environment
- Working Group 4: Transfer Processes and Data for Radiological Impact Assessment
- Working Group 5: Exposure and Effects to Biota
- Working Group 6: Biosphere Modelling for Long Term Safety Assessments of High Level Waste Disposal Facilities
- Working Group 7: Assessment of Fate and Transport of Radionuclides Released in the Marine Environment

The activities and results achieved by the Working Groups are described in individual IAEA Technical Documents (IAEA TECDOCs). This publication describes the work of Working Group 2.

1.2. BACKGROUND FOR MODARIA II WORKING GROUP 2: ASSESSMENT OF EXPOSURES AND COUNTERMEASURES IN URBAN ENVIRONMENTS

The work described in this publication is a natural continuation of that completed by the Urban Remediation Working Group of the EMRAS programme [1–10] and the first phase of the MODARIA programme (2012–2015) by MODARIA Working Group 2 (Assessment of radioactive contamination, exposures and countermeasures in urban environments) and reported in the respective MODARIA Working Group 2 report [9, 11–13]. Since a 'Urban Exposures' Working Group was established in the EMRAS programme, it has had the aim to test and improve the capabilities of models used in assessment of radioactive contamination in urban settings, including short term and long term contaminant redistribution following dispersion and deposition events, as well as those developed to predict the impact of potential countermeasures or remediation efforts for reducing human exposures and doses [7, 12].

1.3. OBJECTIVE

The primary objective of WG2 was to continue the work undertaken during MODARIA I. The key areas of work, similar to those in MODARIA I, were to test and improve the prediction of:

(1) Contamination levels and activity concentrations in air following an atmospheric dispersion and deposition event;
(2) Changes in radionuclide concentrations or external dose rates as a function of location and time;
(3) The most important contributors (e.g. surfaces or exposure pathways) to doses to human in an urban location following a deposition event;
(4) Expected reductions in radionuclide concentrations, dose rates, or doses to humans following the implementation of various countermeasures or remediation efforts.

A specific objective for the Working Group was undertaking six modelling exercises for different types of situation. Analysis of the output from these modelling exercises included comparison of approaches, models, and modelling results for each type of contamination situation. This publication describes each of the modelling exercises, the models used in the exercises, the approaches and parameter selections used by individual participants, and the results of each exercise.

1.4. SCOPE

Working Group 2 developed and carried out six modelling exercises, including five atmospheric dispersion exercises (two short range and three mid-range) and a sixth exercise dealing with a dose assessment in a contaminated urban area.

One short range atmospheric dispersion exercise was a continuation of an exercise based on field tests involving dispersion of a radionuclide by a small amount of explosive undertaken in MODARIA I [11]. This exercise enabled a comparison of model predictions with measurements, as well as intercomparison of predictions. A comparison of two decision support systems was also made for a hypothetical short range modelling exercise.

Two mid-range atmospheric dispersion exercises based on a set of measurements (air concentrations of a tracer) for releases from a power plant were conducted. These exercises used the measurements for the dispersion scenario developed under MODARIA I to explore differences between different modelling approaches; a diagnostic approach (based on measured meteorological information, and a prognostic approach (based on meteorological forecasts). An additional mid-range atmospheric dispersion exercise used a set of measurements (air concentrations of a tracer) of releases from a research reactor to compare prediction of the subsequent transport of the contamination.

The final modelling exercise focused specifically on dose assessment in an urban environment that was contaminated by the accident at the Fukushima Daiichi Nuclear Power Plant in 2011.

This publication describes each modelling exercise, and some conclusions based on the exercises.

Some of the results from the work undertaken by WG2 have already been reported elsewhere [9, 14, 15] but are summarized in this publication for completeness.

1.5. STRUCTURE OF THIS PUBLICATION

Section 1 of this publication provides a brief description of the background of the MODARIA II programme and WG2, the group's objectives, and the scope of its activities. Sections 2–7 describe the modelling exercises, including the scenario description, the models used in the exercise, the modelling results, and explanations for agreement or discrepancies among modellers. Section 2 covers a short range atmospheric dispersion exercise based on a field test, and Section 3 a comparison of decision support systems based on a hypothetical short range atmospheric dispersion situation. Sections 4–6 describe the mid-range atmospheric dispersion exercises, Section 7 describes the dose assessment exercise, and Section 8 provides general conclusions of the Working Group based on the modelling exercises. The Appendix provides additional detailed descriptions of the application of some of the models used in the exercises, whilst a complementary Electronic Appendix provides the meteorological data and monitoring data for the Chalk River modelling exercise which is described in detail in Section 6 of this publication.

2. SHORT RANGE ATMOSPHERIC DISPERSION EXERCISE–BOLETICE

2.1. OVERVIEW AND RATIONALE

The Boletice exercise was a short range atmospheric dispersion exercise based on experimental data obtained from a field test performed by the Czech National Radiation Protection Institute (SÚRO). The field test involved the dispersal of a short lived radionuclide with a small amount of explosive [2, 9, 11, 16]. This exercise provided an opportunity to test model predictions for a short range dispersion event, including the resulting deposition. The exercise presented in this publication is similar to previous exercises [2, 9, 11] but is based on a field test carried out at a different location (Boletice instead of Kamenná) and using a different radionuclide (^{140}La instead of ^{99m}Tc). The Boletice site provided a larger test area, which allowed an omnidirectional dispersion event. Use of ^{140}La (half-life 1.7 d, as compared with 6 h for ^{99m}Tc) permitted a greater number of deposition measurements to be made, at distances farther from the dispersion point, than was the case with the earlier exercises.

The test site is described in Section 2.2, and the experimental conditions for the field test are summarized in Sections 2.3 and 2.4. The radioactive material, a short lived radionuclide (^{140}La) in liquid form, was spread by the detonation of a small amount of explosive in an open field. Input information for the field test provided to the participants included information about the explosion event, the amount of radioactivity involved, the arrangement of the various detectors in the vicinity of the explosion, and meteorological information. The primary endpoint to be modelled for the exercise was surface contamination (Bq/m^2) as a function of distance, although other endpoints (e.g. dose rates, activity concentrations in air) could also be modelled.

The exercise comprised a fully blind model test. Only the input information was provided to participants during the exercise and comparisons were made with measurements only after the modelling results were submitted. The analysis discussed in the following subsections is limited to the predicted and measured surface contamination.

2.2. DESCRIPTION OF THE TEST SITE

The Boletice military training area is located in the southern part of the Czech Republic. The whole training area covers about 220 km^2; the area of interest for the field test covers 400 × 400 m^2; (coordinates in Table 1). The measurement equipment was distributed over an area of about 60 × 60 m^2 (Fig. 1). The test site was covered with grass and other small plants (e.g. clover) and was lined with groups of mostly broad leaved trees or small woods and various bushes. In many cases, the trees were more than 15 m high. A digital elevation model (EU-DEM[1]) was provided to participants in the exercise for terrain visualization and height maps. Relative heights ranged from 0 to 38 m.

[1] https://www.eea.europa.eu/data-and-maps/data/copernicus-land-monitoring-service-eu-dem

TABLE 1. COORDINATES FOR THE CORNERS OF THE TEST AREA, SHOWN BOTH IN UTM COORDINATE SYSTEM (UTM 33N-EPSG:32633) IN METERS AND STANDARD GEOGRAPHIC WGS LONGITUDE/LATITUDE IN DECIMAL DEGREES

Point	X_UTM (m)	Y_UTM (m)	Longitude (deg)	Latitude (deg)
1 (NW)	435784.209387	5416916.223199	14.123761	48.901861
2 (NE)	436184.209387	5416916.223199	14.129219	48.901902
3 (SE)	436184.209387	5416516.223199	14.129281	48.898304
4 (SW)	435784.209387	5416516.223199	14.123824	48.898263

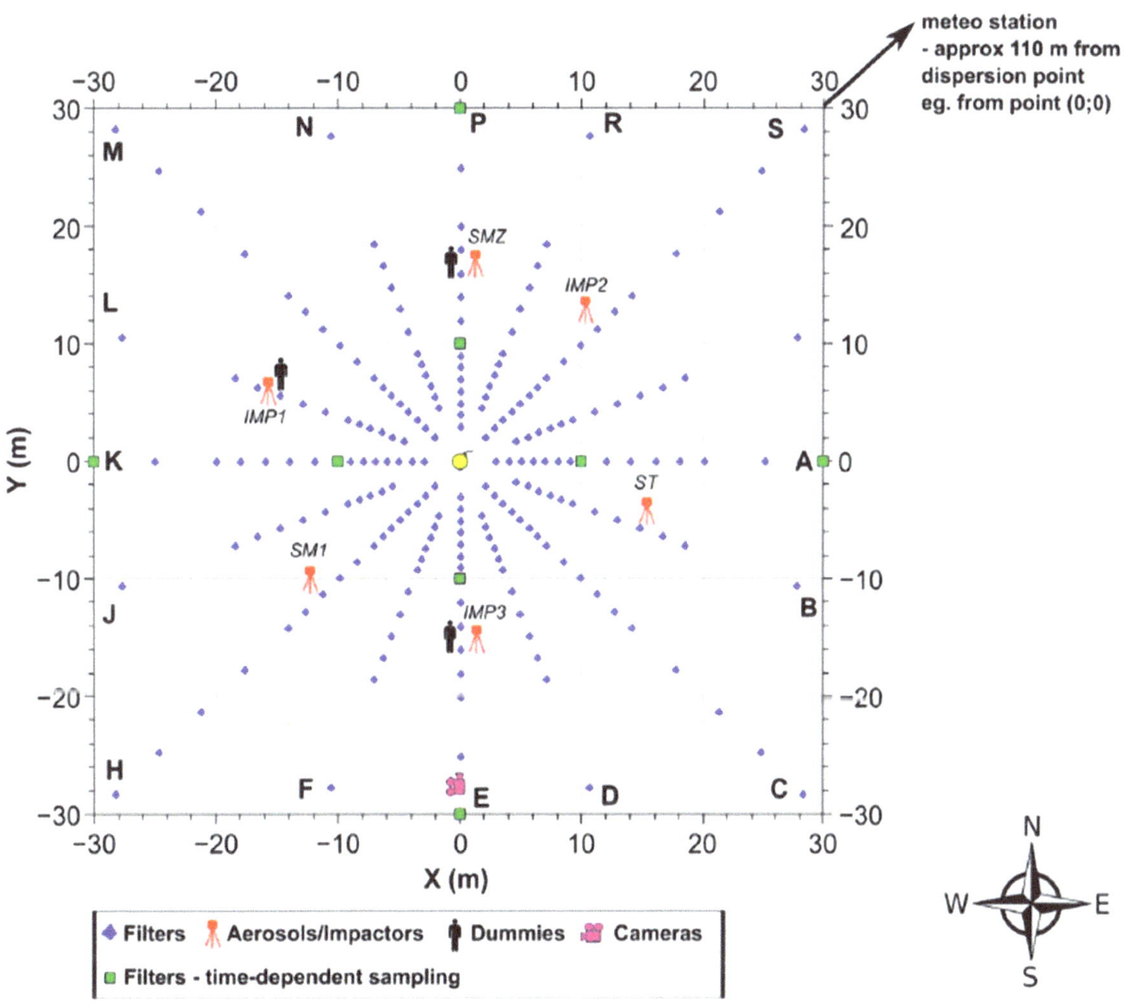

FIG. 1. Locations of the sampling devices for the field test. Aerosol samplers and impactors are labelled ST, SMZ, SM1, IMP1, IMP2, and IMP3. The dummies include a man (bottom), a woman (top), and a child (upper left). NOTE: The letters A–S indicate individual radial lines used to mark the position of the filters and other objects.

TABLE 2. SUMMARY INFORMATION FOR THE BOLETICE FIELD TEST

Date	Explosion time[a]	Radionuclide and activity (MBq)	Amount of liquid containing the activity	Amount and type of explosive used
17 June 2014	17:32	La-140, 713	4 × 10 mL	SEMTEX 1A, 250 g

[a] 24 hour system (12:00 = noon).

2.3. DESCRIPTION OF THE FIELD TEST

The field test was performed by the Czech National Radiation Protection Institute (SÚRO) on a test area within the Boletice military training area. The radioactive material was ^{140}La (half-life 1.7 d) in liquid form ($LaHNO_3$ in an 0.1 M $LaHNO_3$ solution), placed in four 10 mL polyethylene vials; the total activity was 713 MBq. The liquid was spread by detonation of a small amount of explosive (SEMTEX 1A, 250 g) in an open field (flat terrain), in an arrangement that allowed an omnidirectional dispersion. The measurements performed included dose rates, surface contamination of the ground, surface contamination on dummies (man, woman, child) wearing Tyvek overalls, aerosol sampling, and video recording using both standard and high speed cameras. Information about the field test is summarized in Table 2.

2.4. METEOROLOGICAL SITUATION DURING THE FIELD TEST

Meteorological information for the field test is summarized in Table 3. Detailed (time dependent) meteorological information was provided to participants in electronic form. Additional meteorological data from the Temelín station (operated by the Czech HydroMeteorological Institute) were also available to participants. The Temelín station is located 37 km from the test area (longitude, 14.341667; latitude, 49.198333; elevation, 500 m above sea level).

The primary meteorological station was located 110 m to the northeast of the detonation point. This station consisted of a telescopic 10 m high meteorological mast, equipped with an ultrasonic sensor for measuring wind speed and direction. There were additional sensors for air temperature, humidity and pressure at 2 m and 10 m. The measurement interval was 1 s. Close to the mast was a second ultrasonic sensor at 2 m height, also measuring wind speed and direction at an interval of 1 s.

Figure 2 shows the weather conditions during the Boletice field test at the meteorological station located 110 m northeast of the dispersion point. The wind speed immediately after the detonation ranged from about 2 to 8 km/h (0.56–2.2 m/s) at a 2 m height and from about 1 to 7 km/h (0.28–1.9 m/s) at a 10 m height. The wind direction appeared to shift from about 240° at the detonation time to about 270–300° (2 m) or 300–330° (10 m) shortly after the detonation.

2.5. MODELS USED IN THE EXERCISE

Table 4 summarizes the models and parameter values used by participants. The models represent two main types of computational approach to modelling atmospheric dispersion (Gaussian and Lagrangian) and have been developed for various purposes. Three participants provided predictions for this exercise. Descriptions of the individual models used in this exercise are also provided in earlier IAEA publications [2, 11]. Further detailed descriptions of how three of the models were applied to this modelling exercise are given in Appendix I.

TABLE 3. SUMMARY OF WEATHER CONDITIONS DURING THE BOLETICE FIELD TEST[a]

Parameter	Height	Mean	Range
Temperature (°C)	2 m	13.8	13.4–15.4
	10 m	13.5	13.2–15.0
Relative air humidity (%)	2 m	62	52–67
Wind speed	10 m		
(km/h)		2.8	0.07–16.1
(m/s)		0.78	0.019–4.5
Wind direction (deg)	10 m	240.5	0–359
Air pressure (hPa)	2 m	929.8	929.5–930.2

[a] More detailed meteorological data were provided in electronic form. Measurements were taken at 2 m or 10 m height. The indicated wind direction is the direction the wind was blowing from.

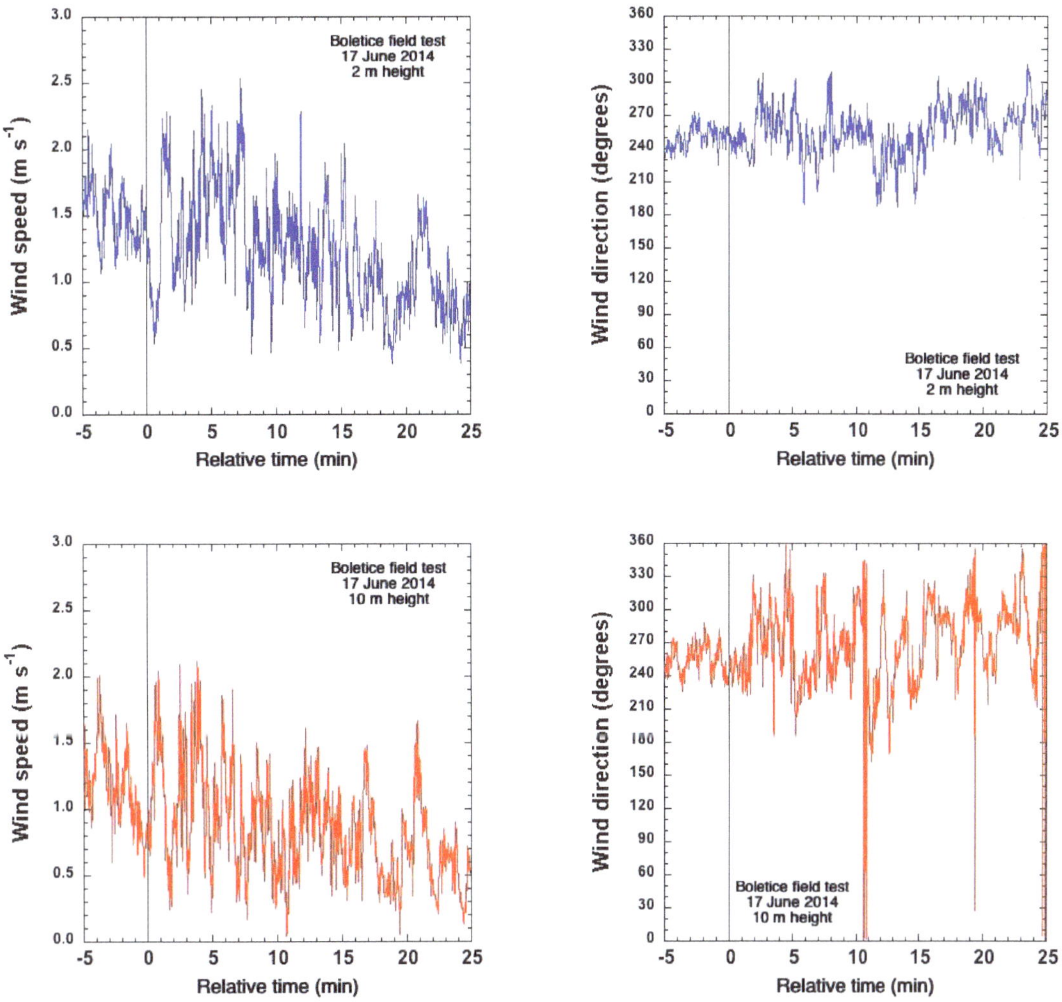

FIG. 2. Wind speed (left) and wind direction (right) for the Boletice field test conducted on 17 June 2014 at 2 m height (top) and 10 m height (bottom), at 1 s intervals. The x-axis represents time relative to the detonation time, indicated by the vertical line at t = 0. The meteorological station was located 110 m northeast of the detonation point (reproduced from Ref. [9] with permission courtesy of Journal of Radiological Protection).

TABLE 4. COMPARISON OF MODELS AND SELECTED PARAMETERS USED IN THE BOLETICE SHORT RANGE ATMOSPHERIC DISPERSION EXERCISE

Model name	**LASAIR v. 4**	**LASAIR 4.0.5**	**HotSpot 3.0.3**	**HotSpot 3.1**
Participant and country	H. Walter, Germany	F. Mancini, Italy	T. Charnock, United Kingdom	F. Mancini, Italy
Type of model	Lagrangian	Lagrangian	Gaussian	Gaussian
Purpose of model	Decision support, emergency response	Emergency response	Emergency response	Emergency response
Number of Lagrangian particles	500 000	500 000	Not applicable	Not applicable
Domain size and grid size	40 km × 40 km or 20 km × 20 km; 5 m grid, increasing to the outside	20 km × 20 km; 5 m × 5 m grid	Nested grid: 5 × 5 m to 100 m downwind; 10 × 10 m to 200 m; 50 × 50 m to 1000 m; 100 × 100 m to 2000 m	Plume centreline only; 10 m × 10 m grid
Handling of meteorological data	Mean value during 1 min intervals (17:32–17:52), starting at 17:34	Mean value during 1 min intervals (17:32–17:41); wind direction shifted by 45 deg	Constant windspeed and direction	Median value during interval 17:32:00–17:32:59
Stability class	D	B	D	B
Wind speed (m/s)	Time dependent, 0.5–1.3	Time dependent, 0.58–1.33	(Run 1)[a] 4 (Run 2) 1.2	1.33
Wind conditions	Transient, 214–298 deg	Transient, 291–326 deg (wind direction shifted by 45 deg)	(1) Steady state, 250 deg (2) Steady state, 286 deg	Steady state, 291 deg
Dry deposition velocity (m/s)	< 2.5 μm, 1×10^{-3} 2.5–10 μm, 1×10^{-2} 10–50 μm, 5×10^{-2} > 50 μm, 1.5×10^{-1} 0–10 μm assumed to be respirable	< 2.5 μm, 1×10^{-3} 2.5–10 μm, 1×10^{-2} 10–50 μm, 5×10^{-2} > 50 μm, 1.5×10^{-1} 0–10 μm assumed to be respirable	Respirable fraction[b], 0.003 Non-respirable fraction, 0.1	Respirable fraction[b], 0.003 Non-respirable fraction, 0.1
Source term partitioning	Homogeneously distributed in the cloud, released within 1 s	Homogeneously distributed in the cloud, released within 1 s	20% at 0.8 height 35% at 0.6 height 25% at 0.4 height 16% at 0.2 height 4% at ground level	20% at 0.8 cloud top 35% at 0.6 cloud top 25% at 0.4 cloud top 16% at 0.2 cloud top 4% at ground level

TABLE 4. (cont.)

Model name	LASAIR v. 4	LASAIR 4.0.5	HotSpot 3.0.3	HotSpot 3.1
Column (cloud) dimensions	Horizontal extension = 12 m Vertical extension = 19 m	Horizontal extension = 12 m Vertical extension = 20 m	Height, 17 m (calculated by Hotspot)	Height, 22 m
Surface roughness	Near (~ 40 × 50 m), 0.01 m Other areas, 1.0 m	0.01 m	0.01 m	0.01 m
Particle size distribution (% of activity per particle size intervals)	< 2.5 µm, 50% < 10.0 µm, 30% < 50.0 µm, 20%	< 2.5 µm, 20% < 10.0 µm, 40% < 50.0 µm, 20% > 50 µm, 20%	Respirable[b], 20% Non-respirable, 80%	< 1 µm, 60% > 1 µm, 40%
Time to set up and run	20 minutes	1 h	5 minutes	10 minutes
Time to process results	5 minutes	5 minutes	5 minutes	10 minutes

[a] The HotSpot 3.0.3 model was run twice by Charnock. The second model run included an adjustment to the timing of the meteorological data to allow for the distance (110 m) between the meteorological station and the dispersion point; all other input parameters and assumptions were the same between the two model runs.

[b] The respirable fraction is the fraction of aerosolized material that is respirable, generally considered as having an Activity Median Aerodynamic Diameter (AMAD) of ≤ 10 µm; the non-respirable fraction is the fraction of aerosolized material that has an AMAD > 10 µm. In HotSpot, the respirable fraction is assumed to have an AMAD of 1 µm [17].

2.6. ANALYSIS OF MODELLING RESULTS

2.6.1. General approach

The analysis of the results focuses on the predicted and measured deposition (surface contamination, Bq/m^2), using the same approach for the comparison of the results developed during the EMRAS II programme [2] and also used during MODARIA I [11]. Deposition profiles were defined by the dispersion point (0,0) and the coordinates of points with predicted or measured deposition (Bq/m^2). The measurements and the model outputs (predictions) from the participants were compared.

Values of activity concentrations were calculated from measurements using a Multilevel B-Spline interpolation [18] method with SAGA GIS[2] software, and the interpolated values were used instead of the measured values [2]. Interpolation of model predictions and measurements using the same method and settings was performed to allow comparisons of contamination densities for the same set of point locations. For the comparisons, it was necessary to put all model outputs into the same coordinate system with the dispersion point at (0,0). The development of the grids was described in detail in the EMRAS II report [2]. The interpolated grids created for each set of model predictions and for the measurements were used as data input for the profiles discussed later (Section 2.6.3).

The plots of the processed data sets (Fig. 3) show the predicted and measured activity concentrations (deposition); the same coordinate system and colour scale are used for different model predictions. The plots therefore enable a visual comparison of the two dimensional predicted or measured contamination and the degree of contamination. The measurements and predictions were normalized to the maximum value of the measured or predicted deposition (1 = the maximum measured or predicted deposition, as relevant). The two plots for HotSpot 3.0.3 as used by Charnock show predictions from two model runs, with the second run including an adjustment to the time dependent meteorology. The predictions by Charnock did not include the first 10 m of the grid area.

The measurements indicated that deposition occurred primarily toward the southeastern part of the grid (Fig. 3). For LASAIR, both sets of predictions predicted a similar direction of the plume; the HotSpot 3.1 results by Mancini (Italy) and the second HotSpot 3.0.3 results by Charnock predicted a similar direction. The first results by Charnock (United Kingdom) predicted a plume toward the northeast; for the second prediction, an adjustment was made to the timing of the meteorological data to account for the distance (110 m) between the meteorological station and the dispersion point; this adjustment resulted in the predicted plume going toward the southeast as with the other model predictions. Thus, the two sets of HotSpot 3.0.3 predictions by Charnock differed only in the wind speed and direction used for the calculations (Table 4). The other model predictions also accounted for the distance between the meteorological station and the dispersion point: Walter (Germany) allowed for a 2 minute difference in the timing of the meteorological data, while Mancini (after observing the videos of the event) revised the reported wind directions by 45°.

[2] http://www.saga-gis.org/saga_module_doc/2.1.3/grid_spline_4.html

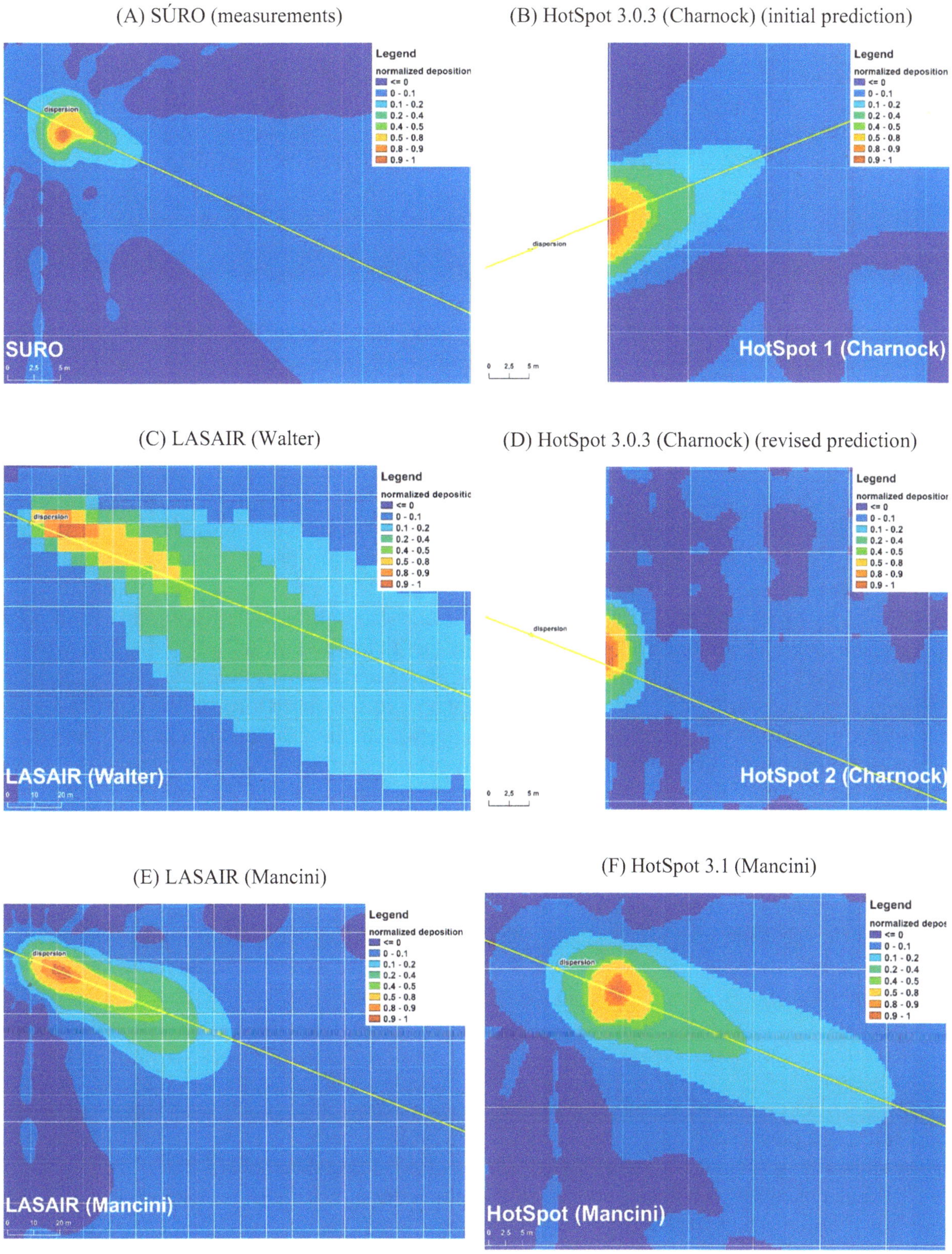

FIG. 3. Contour plots of the measured and predicted deposition for the Boletice field test (17 June 2014). Data are normalized to the maximum value of the measured or predicted deposition (1 = the maximum measured or predicted deposition). The star indicates the dispersion point, and the line indicates the axis of the cloud. Plot A represents the measurements made by SÚRO. Plots B, D, and F show results using two versions of HotSpot (Gaussian model). Plots C and E show results using LASAIR (Lagrangian model). (note that the plots are on different scales). (reproduced from Ref.[9] with permission courtesy of Journal of Radiological Protection).

2.6.2. Maximum activity and total activity in the grid area

Table 5 summarizes the maximum measured and predicted deposition (Bq/m^2, with the coordinates of the location) and the total measured and predicted activity deposited in the grid area (MBq) for the field test. For LASAIR (two users), the total activity deposited within the grid area ranged from 73.8 to 200 MBq (measured, 41.3 MBq); these predicted values exceeded the measured value by factors of 1.8 (Walter) and 4.8 (Mancini). The HotSpot predictions did not include the entire grid area; therefore, the total deposited activity was not calculated.

Predicted values of the maximum deposited activity ranged from 9.0×10^3 to 5.4×10^6 Bq/m^2 (measured, 8.6×10^5 Bq/m^2), i.e. about 1% of the measured value to about a factor of 6 higher. All of the models predicted the location of the maximum deposited activity to be at a greater distance from the dispersion point than the location of the measured maximum value.

2.6.3. Profiles from (0,0) to maximum and along the cloud axis

Profiles of measurements and model predictions were developed in two ways as previously described [2]: (1) from the dispersion point (0,0) through the point with the maximum measured or predicted value of deposited activity (Table 5); and (2) along the measured or predicted cloud axis (Fig. 3). The cloud axis was defined manually and the profile orientation (crossing the 0,0 point) was defined in the same direction. Table 6 provides the predicted or measured profile integrals (profiles of deposited activity) along the cloud axis, both in terms of total activity (Bq) and percentage of total activity. The profiles of predicted deposition in comparison with the measurements are shown in Fig. 4 (profile through the maximum) and Fig. 5 (profile along the cloud axis); in both figures, the normalized profiles are shown on top and the profiles aligned to their maximum values are shown on the bottom. Aligning the profiles to their maximum values facilitates comparison of the distribution curves, starting at the same relative distance of 0 m. The stepped shape of some of the profiles is attributable to differences in resolution between the profile and the grid. Differences in the predicted directions of the profiles are not reflected in Figs 4 and 5.

For the profiles through the cloud axis (Table 6), the total predicted activity (excluding Charnock's first run with HotSpot 3.0.3, before adjustment of the meteorological information) varied over a range of 1.5×10^5 to 1.7×10^6 Bq, that is, from about a factor of 3 below the measured value (LASAIR; Walter) to a factor of 3.7 above the measured value (HotSpot 3.0.3; Charnock's second run, following adjustment of the meteorological information). The predictions by Mancini were within a factor of 1.2 to 1.6 of the measured value.

The distributions (% of activity by distance) for all predictions varied from the measurement results. The measurements indicated that about 81% of the activity under the profile was in the 0 to 11 m range (Table 6). For HotSpot versions (excluding Charnock's first run), about 31% (Mancini) or 39% (Charnock's second run) of the activity was deposited within the first 11 m of the profile, with most of the rest between 11 and 21 m (Charnock's second run) or distributed from 11 to 200 m (Mancini). The profiles of the LASAIR predictions had 46% (Mancini) or 60% (Walter) between 31 and 200 m, with the remainder approximately equally distributed from 0 to 11 m, 11 to 21 m, and 21 to 31 m (Table 6).

TABLE 5. PREDICTED AND MEASURED MAXIMUM VALUES OF DEPOSITED ACTIVITY AND TOTAL ACTIVITY DEPOSITED WITHIN THE GRID AREA FOR THE BOLETICE FIELD TEST OF 17 JUNE 2014[a,b]

Model	Coordinates[c]		Maximum deposited activity (Bq/m^2)	Total activity deposited within the grid area (MBq)
	X	Y		
Measurements (SÚRO)	2	-1.9	8.6×10^5	41.3
Model Predictions				
HotSpot 3.0.3 (1) (Charnock)	10	3.5	9.0×10^3	–[d]
HotSpot 3.0.3 (2) (Charnock)	9.6	-2.5	5.4×10^6	–[d]
HotSpot 3.1 (Mancini)	9	-3	3.6×10^5	–[d]
LASAIR (Walter)	12.29	-2.72	2.3×10^4	73.8
LASAIR (Mancini)	12.5	-5	1.1×10^5	200

[a] (reproduced from Ref. [9] with permission courtesy of Journal of Radiological Protection).
[b] The total dispersed activity for the Boletice field test of 17 June 2014 was 713 MBq of ^{140}La.
[c] Coordinates for the locations of the maximum predicted and measured activities, assuming a dispersion point (origin of the explosion) at (0,0); distances are in m.
[d] Not calculated. Some parts of the grid area were not included in the HotSpot predictions.

TABLE 6. INTEGRALS OF DEPOSITION ALONG THE PROFILE THROUGH THE CLOUD AXIS[a], SHOWN BOTH AS TOTAL ACTIVITY (Bq) AND AS PERCENTAGES OF TOTAL ACTIVITY (%)

Relative distance (m)	Measurements (SÚRO)	HotSpot 3.0.3 (1) (Charnock)	HotSpot 3.0.3 (2) (Charnock)	HotSpot 3.1 (Mancini)	LASAIR (Walter)	LASAIR (Mancini)
			Total activity (Bq)			
-10 to 0	30 719	0[b]	0[b]	18 121	1992	21 521
0 to +11	363 548	538	650 129	223 592	17 277	97 170
+11 to +21	18 742	4847	976 742	177 447	22 431	103 679
+21 to +31	10 524	1331	32 452	83 007	17 978	76 606
+31 to +200	26 310	1780	14 008	209 293	87 503	253 482
Total	449 844	8496	1 673 330	711 460	147 182	552 458
			% of total activity			
-10 to 0	6.8	0	0	2.6	1.4	3.9
0 to +11	80.8	6.3	38.9	31.4	11.7	17.6
+11 to +21	4.2	57.1	58.4	24.9	15.2	18.8
+21 to +31	2.3	15.7	1.9	11.7	12.2	13.9
+31 to +200	5.9	21.0	0.8	29.4	59.5	45.9
Total	100	100	100	100	100	100

[a] Calculated for a profile 1 cm in width.
[b] Data not calculated within 10 m of the dispersion point along the x axis. The profile is diagonally oriented, so there are some data within the '0 to +11' m segment along the cloud axis profile.

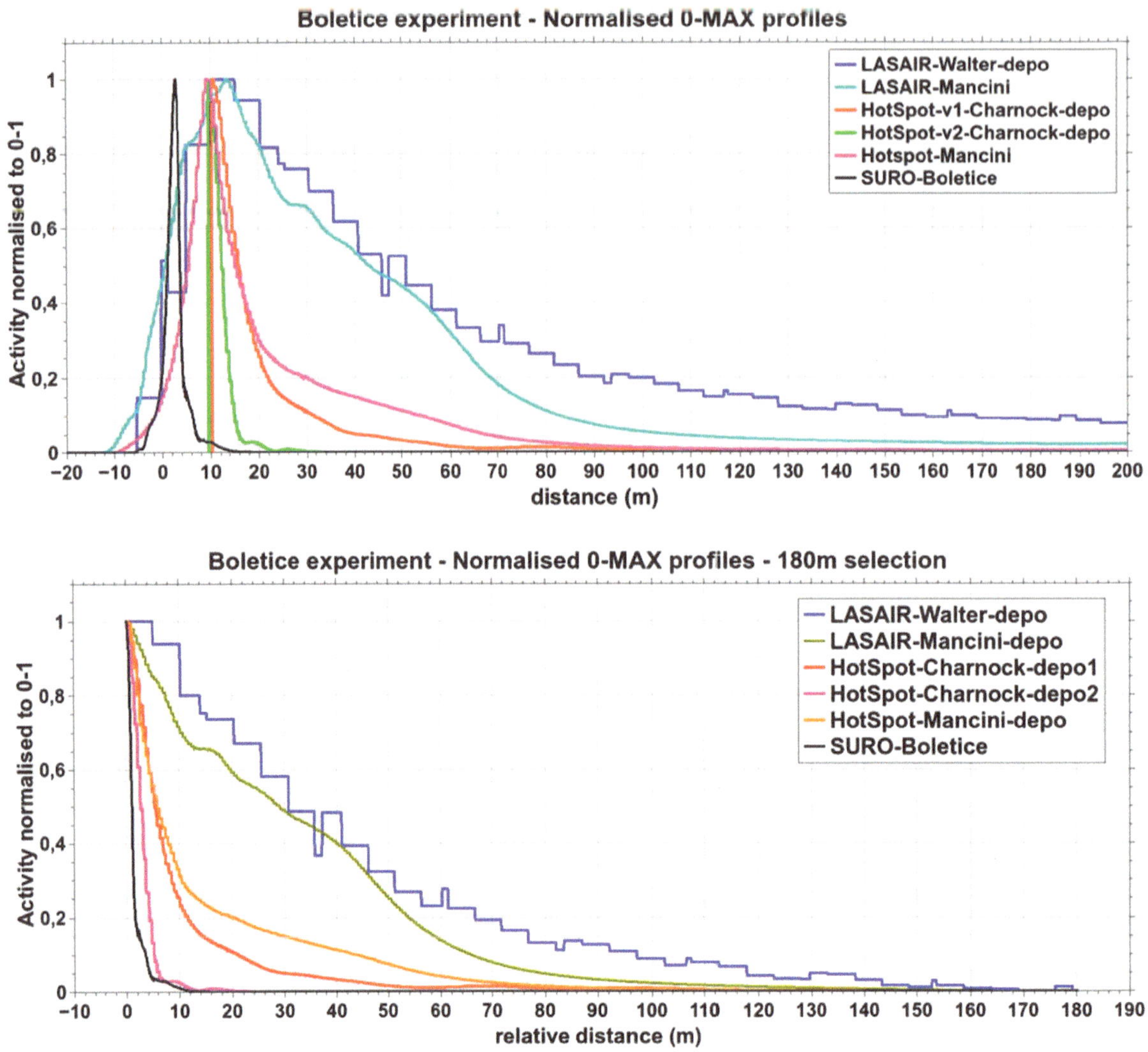

FIG. 4. Profiles (0 to maximum) of the predicted deposition in comparison with the measurements for the Boletice field test. Shown are the normalized profiles with respect to distance (top) and relative distance (aligned at the maximum values, bottom).

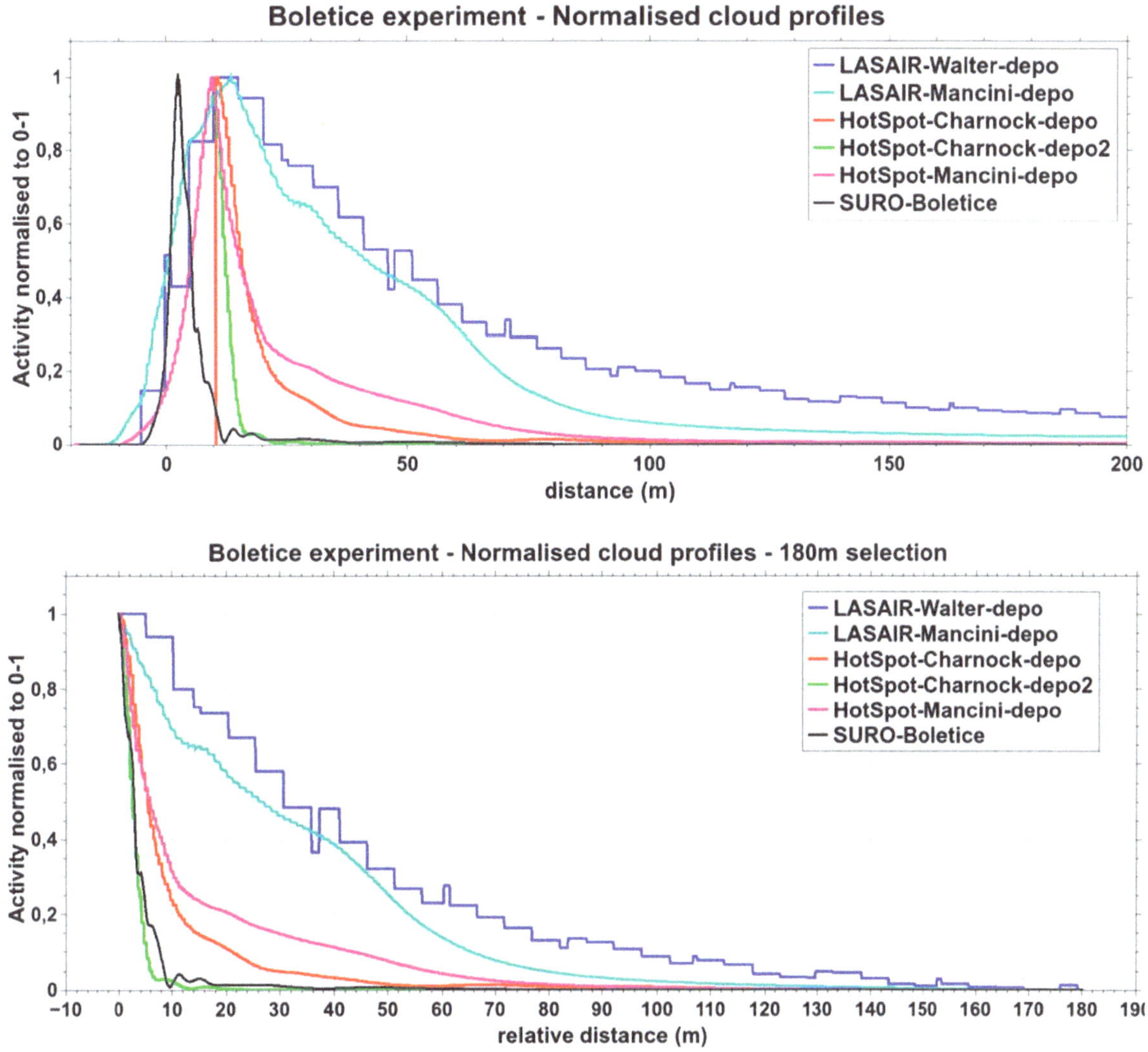

FIG. 5. Profiles (along the cloud axis) of the predicted deposition in comparison with the measurements for the Boletice field test. Shown are the normalized profiles with respect to distance (top) and relative distance (aligned at the maximum values, bottom).

2.7. CONCLUSIONS FROM THE BOLETICE EXERCISE

The Boletice exercise enabled a comparison of model predictions with measurements for an omnidirectional dispersion event, in contrast to the directed dispersion events of the previous field tests (EMRAS II and MODARIA I exercises [2, 9, 11]). The exercise included results from two models and three participants; two participants used the same model (Walter and Mancini with LASAIR), two participants used different versions of the same model (Charnock and Mancini with HotSpot 3.0.3 and 3.1, respectively), and one participant used two models (Mancini with HotSpot 3.1 and LASAIR). For the two versions of HotSpot (3.0.3 and 3.1), the participants made different selections for atmospheric stability class, particle size distribution, and height of the cloud top (Table 4). However, Charnock's second prediction used a wind speed and direction similar to Mancini. Similarly, for LASAIR, participants made different selections for atmospheric stability class, range of wind directions, and particle size distribution (Table 4).

The findings on the exercise are given in Ref. [9] and are summarized here. Based on observations from the videos of the field test, Mancini manually shifted the wind directions by 45° for both the HotSpot 3.1 and LASAIR models. Walter and Charnock also adjusted the wind data, Walter by adjusting the time dependent meteorological data by 2 minutes, and Charnock by selecting average values for wind speed and direction that were consistent with the observed direction of the plume. In earlier exercises involving field tests [2, 9, 11], the meteorological stations were located 20 m or less from the dispersion point, in contrast to 110 m in this exercise. All three participants compensated for this distance, either by delaying time dependent data (Walter, Charnock) or by adjusting the wind direction to match observations from videos (Mancini). The different assumptions made by the participants illustrates the importance of obtaining meteorological data as close as possible to the dispersion point, or accounting appropriately for the distance between them. Lack of an onsite meteorological station, e.g. for an unplanned dispersion event with only regional meteorological data available, would mean even greater difficulty in modelling this kind of dispersion event.

The predicted locations for the maximum deposited activities (Table 5) are similar for the users of HotSpot and for both users of LASAIR. The predicted distributions of the profiles through the cloud axis (Table 6) show corresponding differences between HotSpot and LASAIR. However, for each model, the magnitudes of the predicted maximum deposited activities varied between the two models, HotSpot and LASAIR and between users (as shown in Fig. 4). Mancini's predictions using the two different models differed from each other by a factor of about 3, consistent with the same input assumptions being used for both models.

As in the previous exercises [2, 9, 11], modelling this field test presented the challenge of predicting spatially varying deposition (surface activity concentration) using time dependent meteorological conditions. Using the same coordinate system and colour scale for visual comparison of contour plots of measured and predicted deposition, is helpful. Comparison of modelling results among participants and with the measurements provided the opportunity to better understand the situation being modelled and to improve the application of the models to this specific dispersion event.

3. HYPOTHETICAL DISPERSION EXERCISE IN AN URBAN AREA IN MUNICH

3.1. INTRODUCTION

During MODARIA I, a comparison of two decision support systems (CERES®CBRN-E and LASAIR) was conducted in order to identify more about the similarities and differences between these tools [11]. That comparison was based on a hypothetical dispersion event located in part of Paris, France. The present exercise continues the comparison of CERES®CBRN-E and LASAIR, based on a hypothetical dispersion event located in part of Munich, Germany, in addition to some very simple hypothetical dispersion events.

3.2. DESCRIPTION OF THE EXERCISE

Two initial, very simple scenarios were defined, the first involving only one building, and the second involving two buildings with a street canyon (Fig. 6). The third, more complex scenario was defined in a dense urban area (Munich, Germany), in terms of a building on the corner of Zentnerstrasse and Agnesstrasse (GE Coordinates, 48° 09’ 29.51” N, 11° 233’ 52.58” E (the southwest corner of the building) (Figs 6 and 7). For each scenario, the dispersion of 2.3×10^9 Bq of a radioactive tracer was assumed. Two types of release were considered for each scenario, i.e. a continuous release from a height of 1 m (duration, 10 min), and an explosion of 200 g TNT equivalent. The details of each release scenario are summarized in Table 7 and Figs 6 and 7.

3.3. MODELS USED IN THE EXERCISE

The two decision support systems used in this exercise (CERES®CBRN-E and LASAIR) are described in detail in Ref. [11]. Both decision support systems are intended to provide rapid assessments of atmospheric dispersion in emergency situations. Computationally, for the present context, both systems use Lagrangian particle models to simulate atmospheric dispersion of radionuclides in an urban environment, including the effects of complex terrain such as buildings and street canyons. (The Lagrangian particle model in CERES®CBRN-E is called PMSS, for Parallel Micro Swift Spray; some of the results shown are labelled PMSS in the figures.) Table 8 summarizes the two models and selected parameters used for this exercise.

3.4. ANALYSIS OF MODELLING RESULTS

3.4.1. Scenario 1 (very simple, one building)

Scenario 1 involved a release point upwind of a single building, as shown in Fig. 6(a). The modelling results are shown in Fig. 8 for the continuous release and Fig. 9 for the explosion. For the continuous release (Fig. 8), the contour plots from LASAIR and CERES®CBRN-E look very similar, with LASAIR predicting slightly higher activity concentrations in air than CERES®CBRN-E, but slightly lower deposition. Down the plume centre line (bottom graphs), both models predicted a spike within the first 50 metres (for both endpoints), followed by a gradual decline with increasing downwind distance. LASAIR showed a steeper decline in air concentration than CERES®CBRN-E out to about 250 m, with similar declines thereafter. For the explosion, the predicted magnitudes for activity concentration in air and deposition (Fig. 9) were similar to those for the continuous release, with a greater difference in predicted deposition between the two models and with similar rates of decline with increasing distance for both activity concentration in air and deposition.

TABLE 7. SUMMARY OF MODELLING SCENARIOS FOR THE MUNICH EXERCISE

Scenario	Scenario 1 Very simple	Scenario 2 Simple	Scenario 3 Complex
Description	One building (Fig. 6(a))	Two buildings with urban canyon, two wind directions (Fig. 6(b))	Complex building, Munich, corner of Zentnerstrasse and Agnesstrasse (Figs 6(c), 7)
Wind direction	270°	wind 1, 270° wind 2, 225°	270°
		All scenarios	
Source term (activity dispersed)		2.3×10^9 Bq	
Release conditions: (a) Continuous release (b) Explosion		(a) height = 1 m (1 m × 1 m × 1 m); duration = 10 min (b) 200 g TNT equivalent	
Wind speed at 10 m height		1 m/s	
Atmospheric stability class (Pasquill/Gifford)		D (neutral)	
Roughness length, z_0		0.01 m	

TABLE 8. COMPARISON OF MODELS AND SELECTED PARAMETERS USED IN THE URBAN DISPERSION EXERCISE

Model name	LASAIR v. 4	CERES®CBRN-E (PMSS)
Participant and country	H. Walter, Germany	L. Patryl, France
Type of model	Lagrangian	Lagrangian
Purpose of model	Emergency response	Emergency response
Number of Lagrangian particles	500 000	50 000
Domain size	40 km × 40 km For results of this scenario, the domain size is 1280 m × 1280 m	1800 m × 1800 m (wind 225°) 1800 m × 800 m (wind 270°) 20 vertical levels up to 1500 m
Grid size	5 m resolution in x and y	5 m resolution in x and y
Method of handling meteorological data	Time dependent	Time dependent
Wind field models (level of detail)	Mass consistent wind field model (lprwnd) with consideration of different stability classes	Mass consistent wind field model. Two turbulence schemes: mixing length (local) + Hanna scheme (background)
Dry deposition velocity (m/s)	0–2.5 µm, 1.0×10^{-3} 2.5–10 µm, 1.0×10^{-2} 10–50 µm, 5.0×10^{-2}	0–2.5 µm, 1.0×10^{-3} 2.5–10 µm, 1.0×10^{-2} 10–50 µm, 5.0×10^{-2}
Cloud dimensions (explosion)	Height = 16 m Width = 10 m Length = 10 m	Height = 16 m Width = 3m Length = 3 m
Aerosol (particle size) distribution (%)	0–2.5 µm, 25% 2.5–10 µm, 25% 10–50 µm, 50%	1.25 µm, 25% 6.25 µm, 25% 10 µm, 50%
Building effects	Yes	Yes
Time to set up and run	~10 minutes	~10 minutes
Time to process results	~5 minutes	~5 minutes

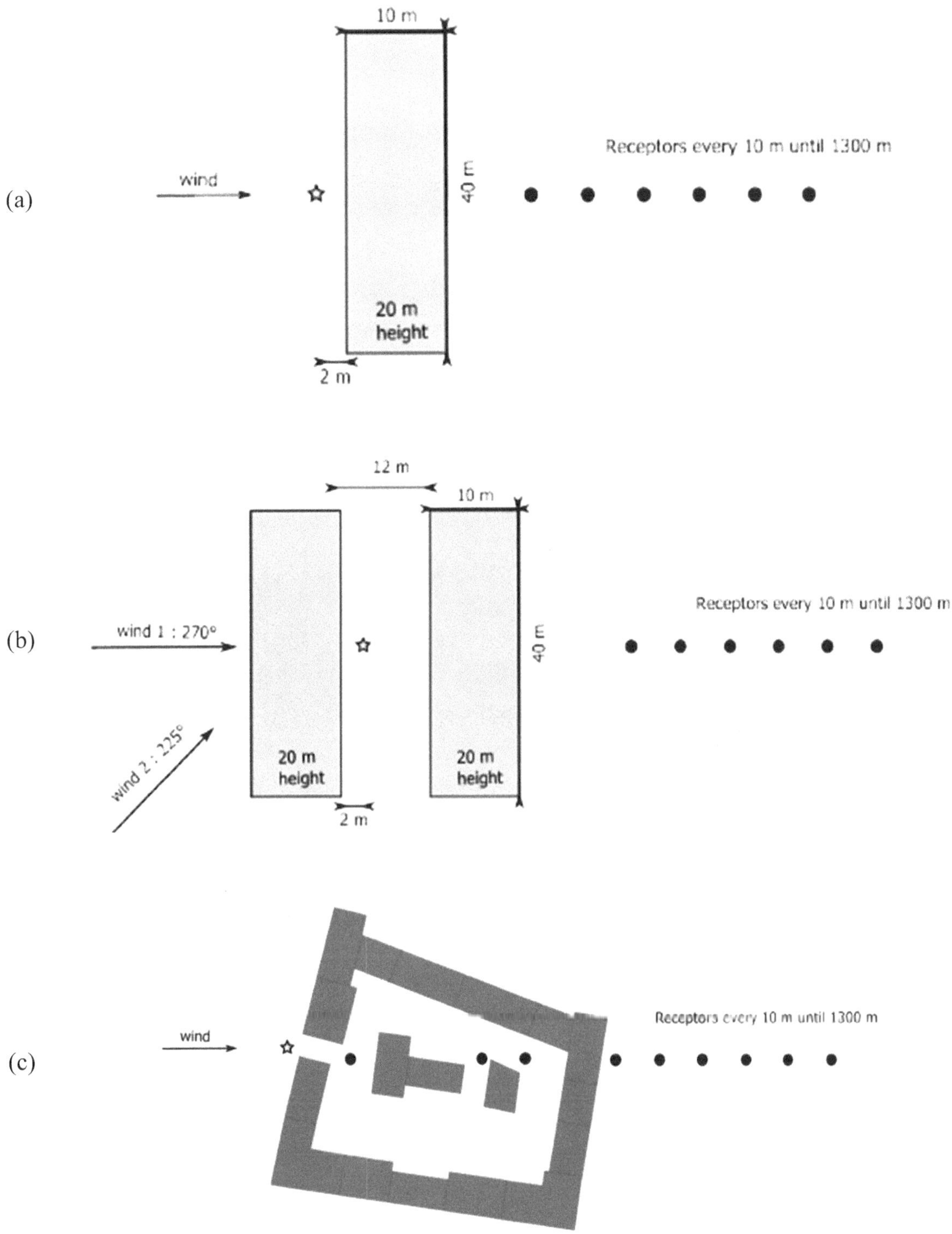

FIG. 6. Summary of modelling scenarios. (a) Scenario 1, very simple, one building; (b) Scenario 2, simple, two buildings, two wind directions; (c) Scenario 3, complex (Munich, see Fig. 7). The star indicates the dispersion point for a given scenario.

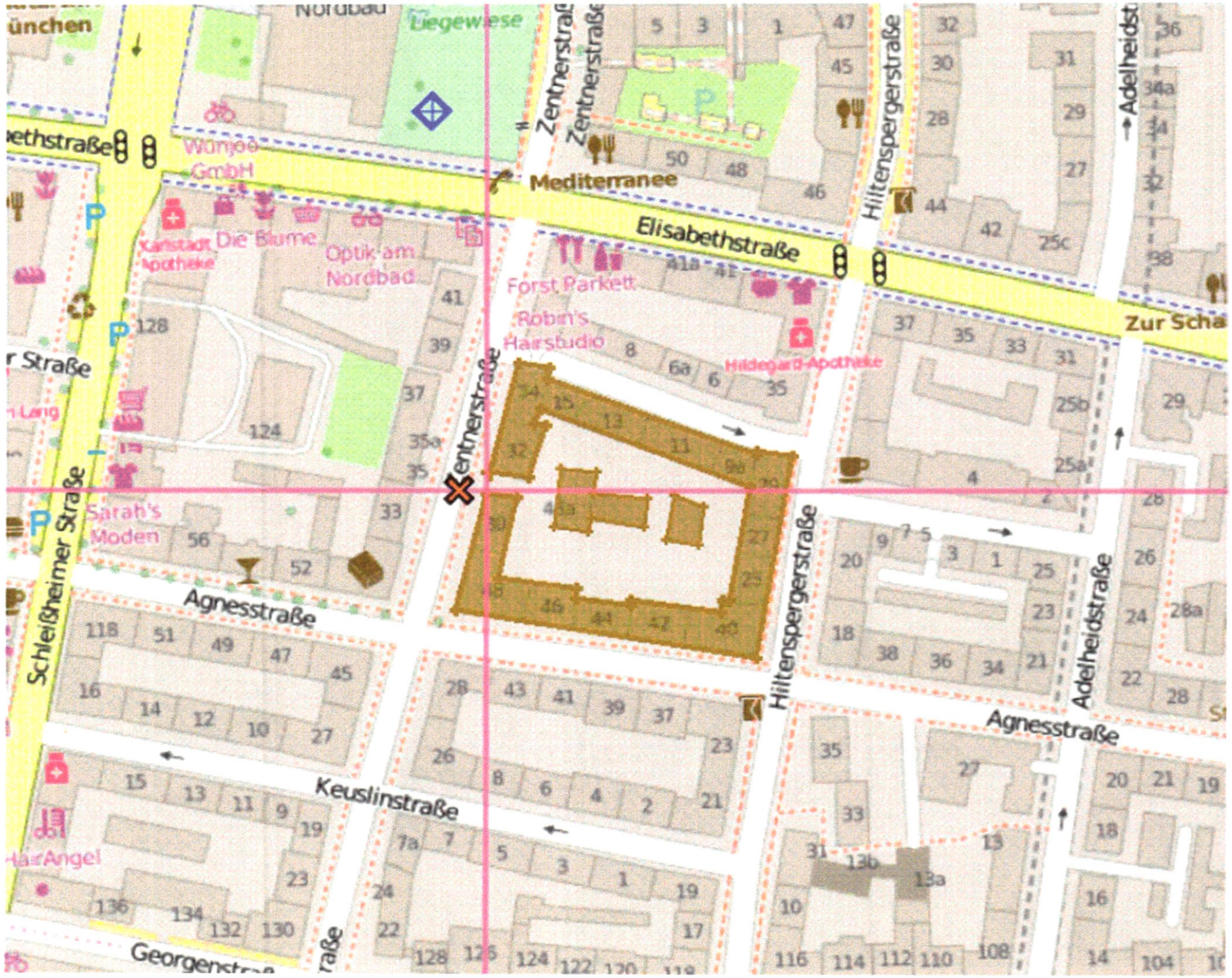

FIG. 7. Map showing the location of the complex building in Scenario 3 (Munich, corner of Zentnerstrasse and Agnesstrasse). The X indicates the dispersion point. (Based on map data from OpenStreetMap https://www.openstreetmap.org/copyright, as used in the model LASAIR.)

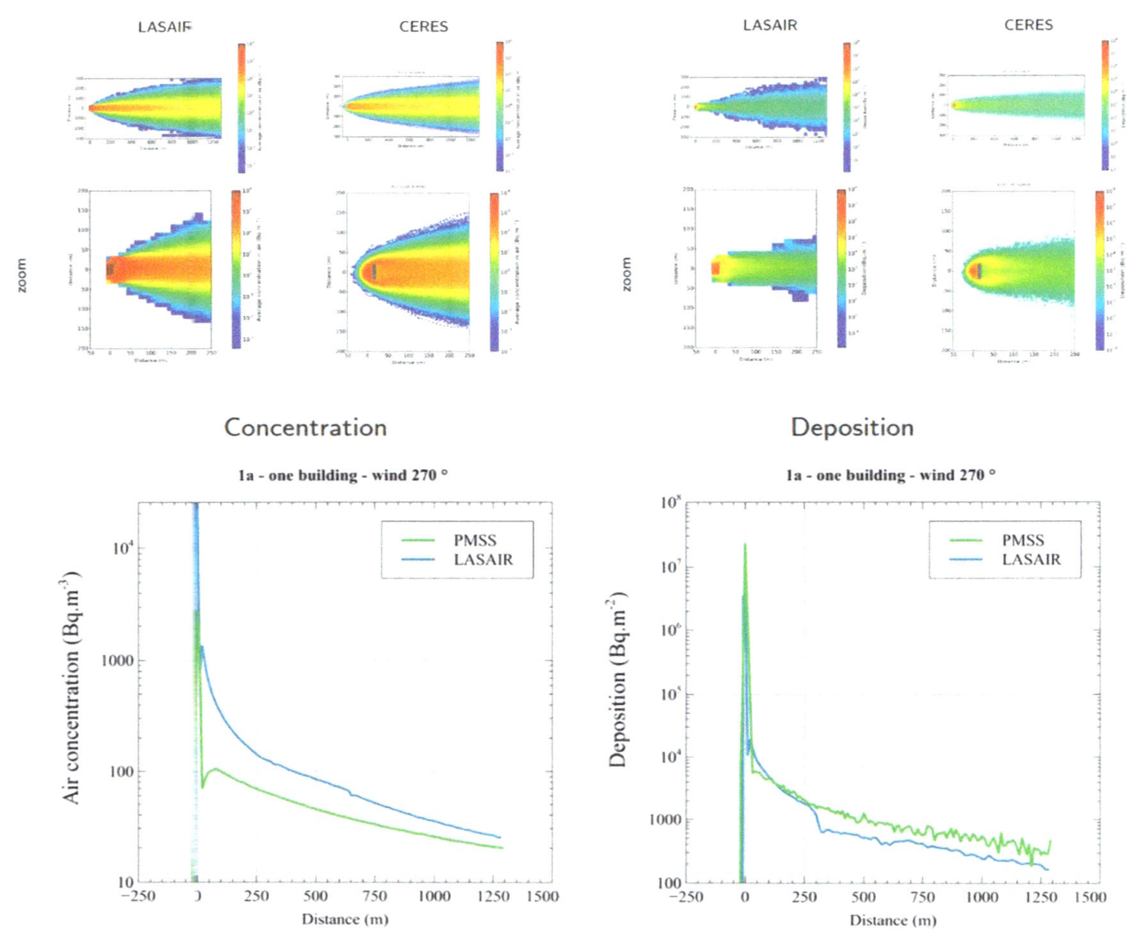

FIG. 8. Modelling results for Scenario 1a (one building, continuous release, wind 270°). The graphs on the left show predicted activity concentrations in air (Bq/m^3, averaged over 1 hour); the graphs on the right show predicted deposition after 1 hour (Bq/m^2). The first row of contour plots shows the predictions out to 1300 m downwind of the release point; the second row ('zoom') shows the predictions over the first 250 m downwind. The bottom graphs compare the predicted activity concentrations in air (left) or predicted deposition (right), as a function of downwind distance along the centre of the plume. (Both 'CERES' and 'PMSS' refer to predictions from CERES®CBRN-E.)

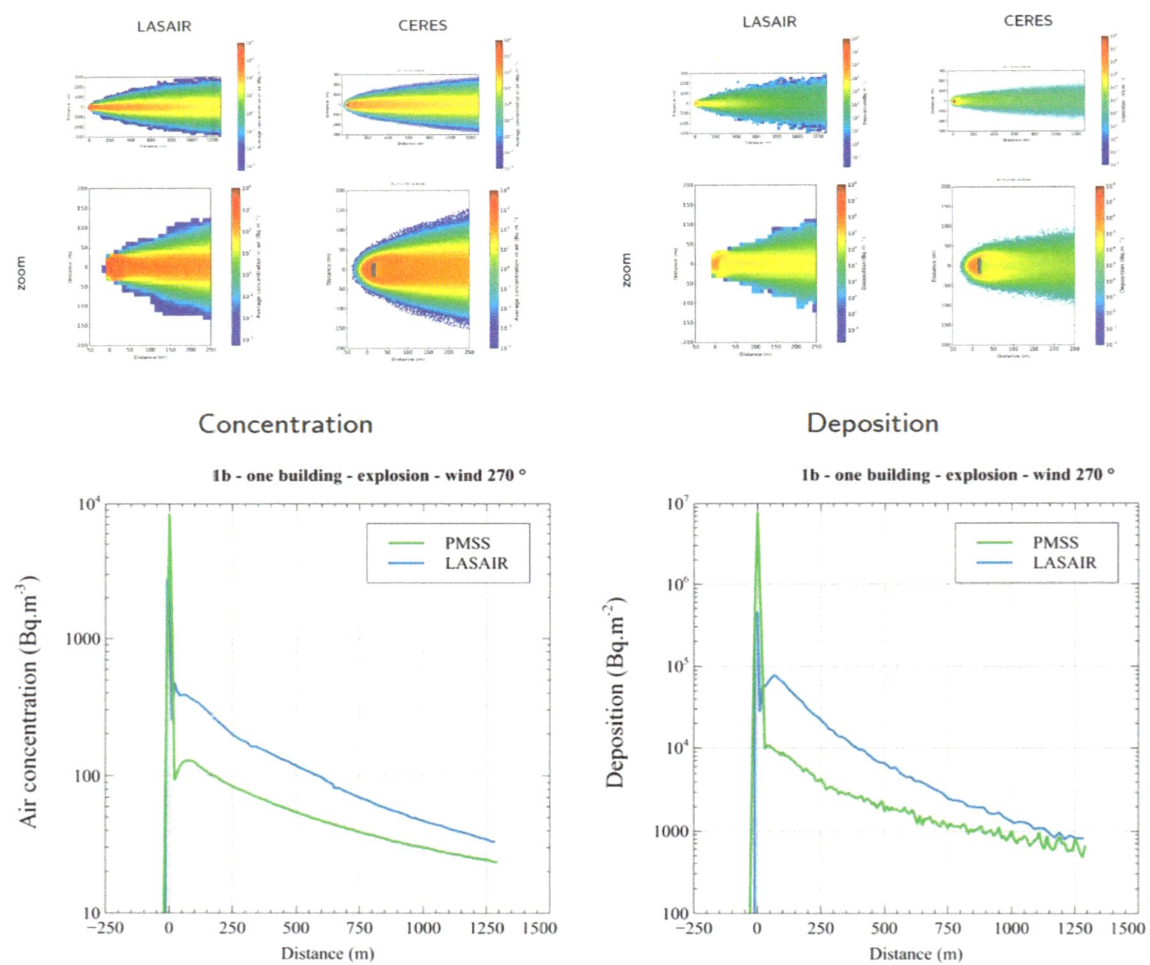

FIG. 9. Modelling results for Scenario 1b (one building, explosion, wind 270°). The graphs on the left show predicted activity concentrations in air (Bq/m^3, averaged over 1 hour); the graphs on the right show predicted deposition after 1 hour (Bq/m^2). The first row of contour plots shows the predictions out to 1300 m downwind of the release point; the second row ('zoom') shows the predictions over the first 250 m downwind. The bottom graphs compare the predicted activity concentrations in air (left) or predicted deposition (right), as a function of downwind distance along the centre of the plume. (Both 'CERES' and 'PMSS' refer to predictions from CERES®CBRN-E.)

3.4.2. Scenario 2 (simple, two buildings), wind 270°

Scenario 2 involved a release point between two buildings, in the urban canyon between them, as shown in Fig. 6(b). This scenario was carried out with two different wind directions. For wind at 270°, for the continuous release, both models again showed an initial spike for both activity concentration in air and deposition (Fig. 10). For both endpoints, LASAIR predicted a sharp decline followed by an increase and then a gradual decline. CERES®CBRN-E showed a very rapid decline in air concentration without a subsequent increase before changing to a gradual decline, while for deposition there was a very steep decline followed by a several orders of magnitude increase, probably corresponding to a 'shadow' due to the building geometry. This 'shadow' was not evident in the predictions for the explosion (Fig. 11), with both the activity concentration in air and deposition predicted by CERES®CBRN-E showing the same spike followed by a gradual decline. LASAIR again showed an increase in both endpoints following the initial spike, and then a gradual decline.

3.4.3. Scenario 2 (simple, two buildings), wind 225°

The second part of Scenario 2 again involved a release point between two buildings, but with the wind at 225°. The results are shown in Fig. 12 for the continuous release and Fig. 13 for the explosion. The difference in wind direction is clearly seen in the contour plots (compared with Figs 10 and 11). The predicted activity concentrations in air and deposition down the plume centreline for the two models were much closer in magnitude for this wind direction than for the wind at 270°. LASAIR did not predict a rise following the initial spike as it did for the wind at 270°, but CERES®CBRN-E did, especially for air concentration. Similar results were obtained for the explosion (Fig. 13), with the two models giving practically identical results between about 50 m and either 650 m (air concentration) or 400 m (deposition). Thus, it seems at least for these building dimensions and configuration, that the effect of the buildings on dispersion was less with the wind at 225° than with the wind at 270°.

3.4.4. Scenario 3 (complex building)

Scenario 3 involved a complex building with a courtyard and small buildings inside the courtyard (Figs 6(c) and 7), and with the face of the building not quite perpendicular to the wind as was the case in the first two scenarios. The modelling results are shown in Fig. 14 for the continuous release and Fig. 15 for the explosion. The model predictions for activity concentration in air are considerably more complex for the first 250 m downwind from the dispersion point for both release types, with a series of spikes, differing between the two models. For deposition, the model predictions looked more similar, with an initial broad spike (higher for CERES®CBRN-E than for LASAIR), followed by a rise and then gradual decline, with LASAIR having greater magnitudes and a steeper decline. Looking at the contour plots, both models give similar activity concentrations in air inside the courtyard. The predictions with CERES®CBRN-E show fluxes both inside and around the building, while the predictions with LASAIR do not seem to account for dispersion around the building.

Figure 16 shows predictions (contour plots) with LASAIR for the complex building and a continuous release (1 min), with the dispersion point at different locations, outside and inside the courtyard. The effect of the dispersion point location is most important close to the building, while downwind the plumes are similar. Thus, the influence of the building may be limited to the vicinity of the building for the conditions modelled; with higher wind speeds and stable meteorological conditions, the area of building influence could extend further downwind. Distances further downwind could be more important for public exposures, as access to the vicinity of the dispersion point could be restricted.

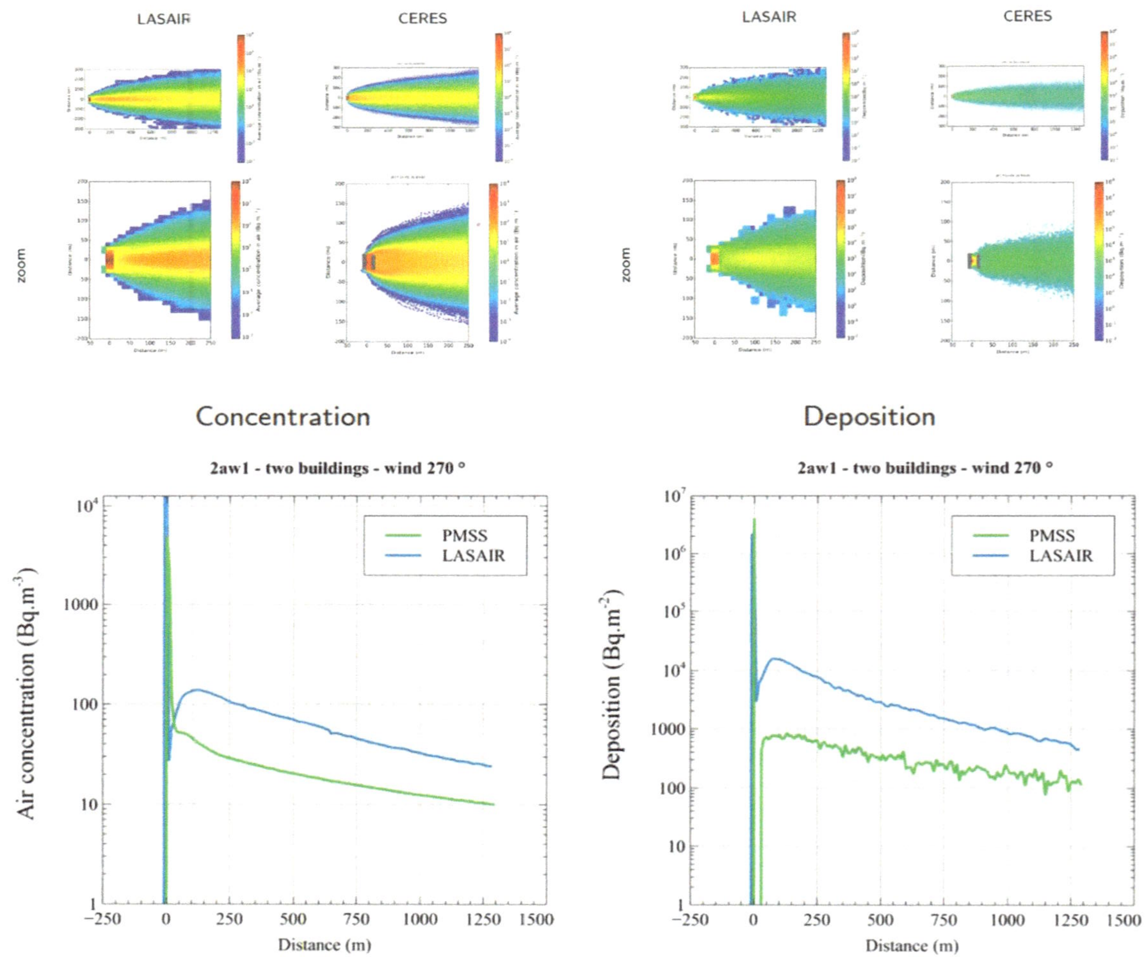

FIG. 10. Modelling results for Scenario 2aw1 (two buildings, continuous release, wind 270°). The graphs on the left show predicted activity concentrations in air (Bq/m^3, averaged over 1 hour); the graphs on the right show predicted deposition after 1 hour (Bq/m^2). The first row of contour plots shows the predictions out to 1300 m downwind of the release point; the second row ('zoom') shows the predictions over the first 250 m downwind. The bottom graphs compare the predicted activity concentrations in air (left) or predicted deposition (right), as a function of downwind distance along the centre of the plume. (Both 'CERES' and 'PMSS' refer to predictions from CERES®CBRN-E.)

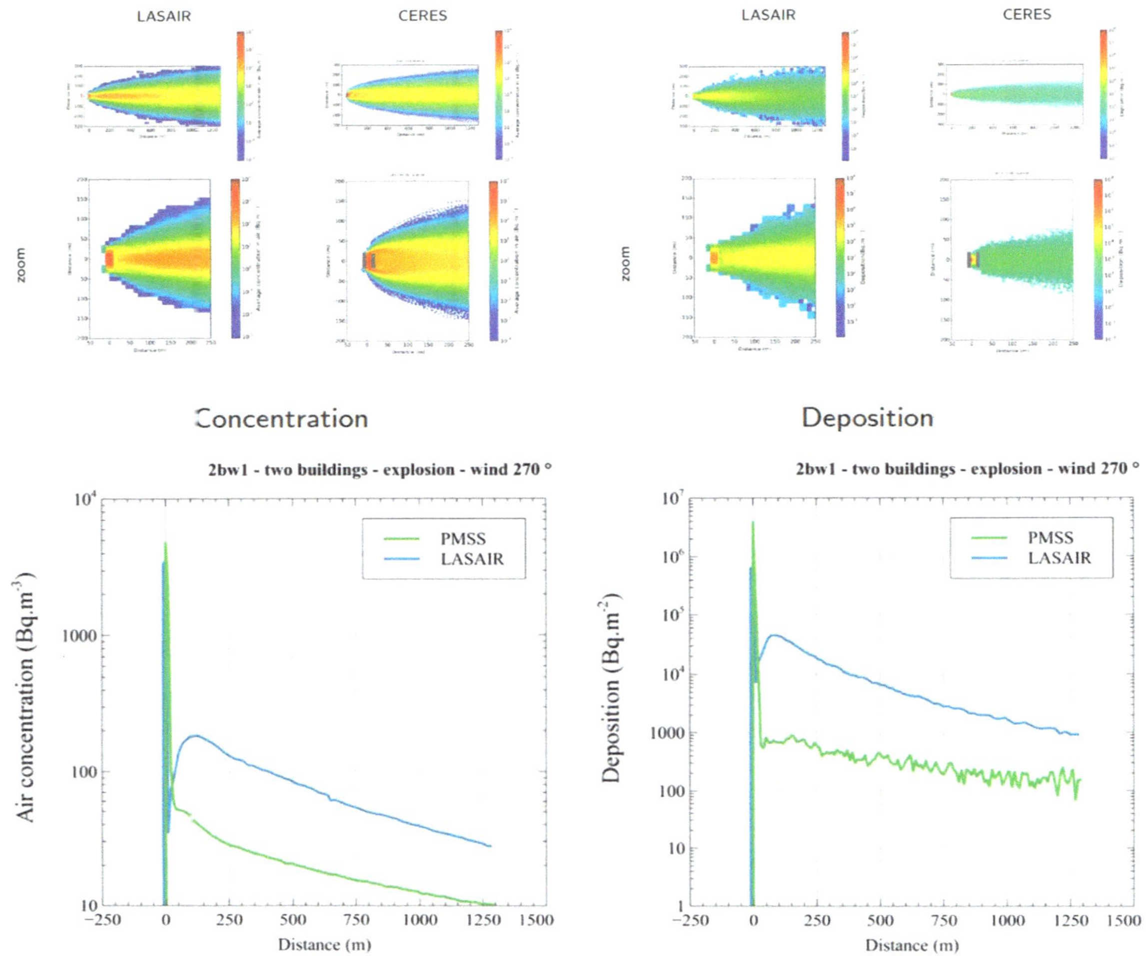

FIG. 11. Modelling results for Scenario 2bw1 (two buildings, explosion, wind 270°). The graphs on the left show predicted activity concentrations in air (Bq/m^3, averaged over 1 hour); the graphs on the right show predicted deposition after 1 hour (Bq/m^2). The first row of contour plots shows the predictions out to 1300 m downwind of the release point; the second row ('zoom') shows the predictions over the first 250 m downwind. The bottom graphs compare the predicted activity concentrations in air (left) or predicted deposition (right), as a function of downwind distance along the centre of the plume. (Both 'CERES' and 'PMSS' refer to predictions from CERES®CBRN-E.)

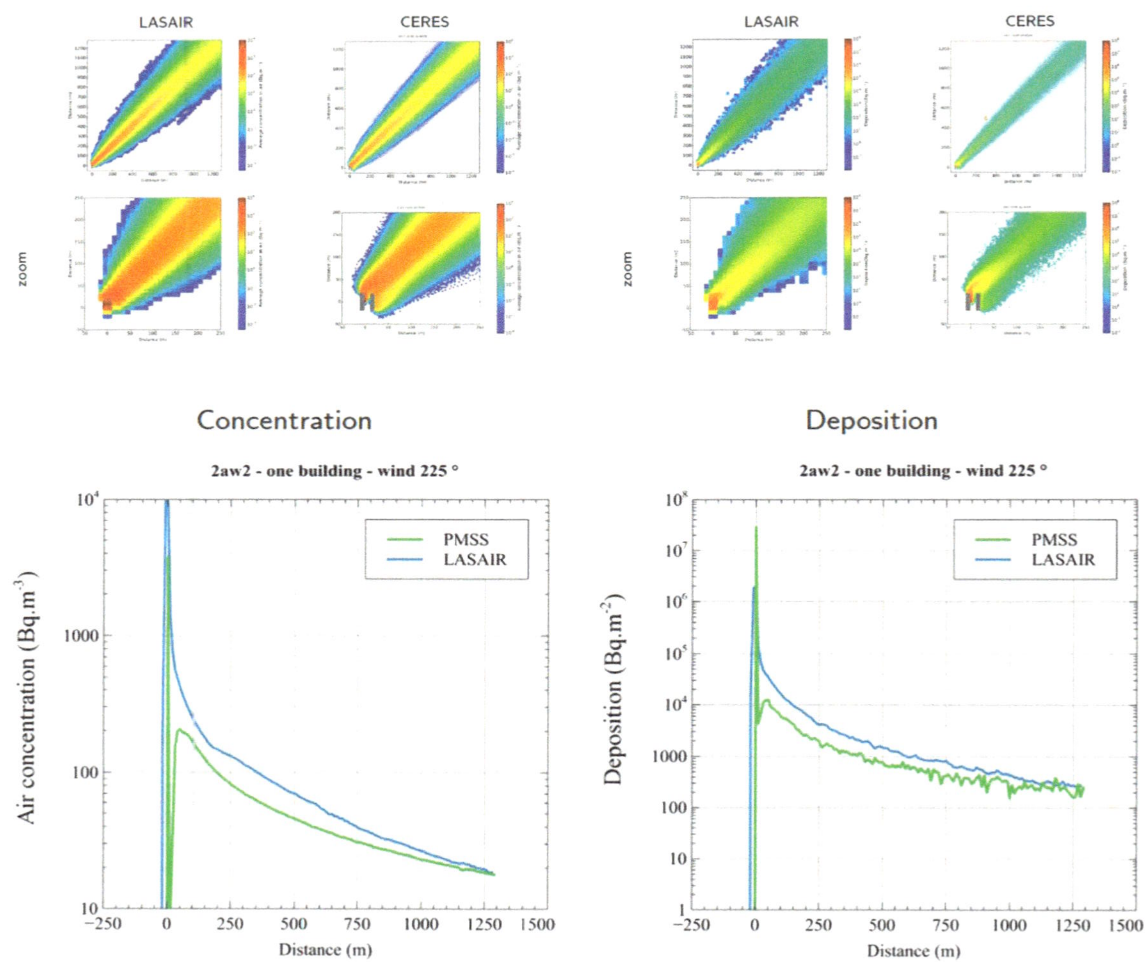

FIG. 12. Modelling results for Scenario 2aw2 (two buildings, continuous release, wind 225°). The graphs on the left show predicted activity concentrations in air (Bq/m^3, averaged over 1 hour); the graphs on the right show predicted deposition after 1 hour (Bq/m^2). The first row of contour plots shows the predictions out to 1300 m downwind of the release point; the second row ('zoom') shows the predictions over the first 250 m downwind. The bottom graphs compare the predicted activity concentrations in air (left) or predicted deposition (right), as a function of downwind distance along the centre of the plume. (Both 'CERES' and 'PMSS' refer to predictions from CERES®CBRN-E.)

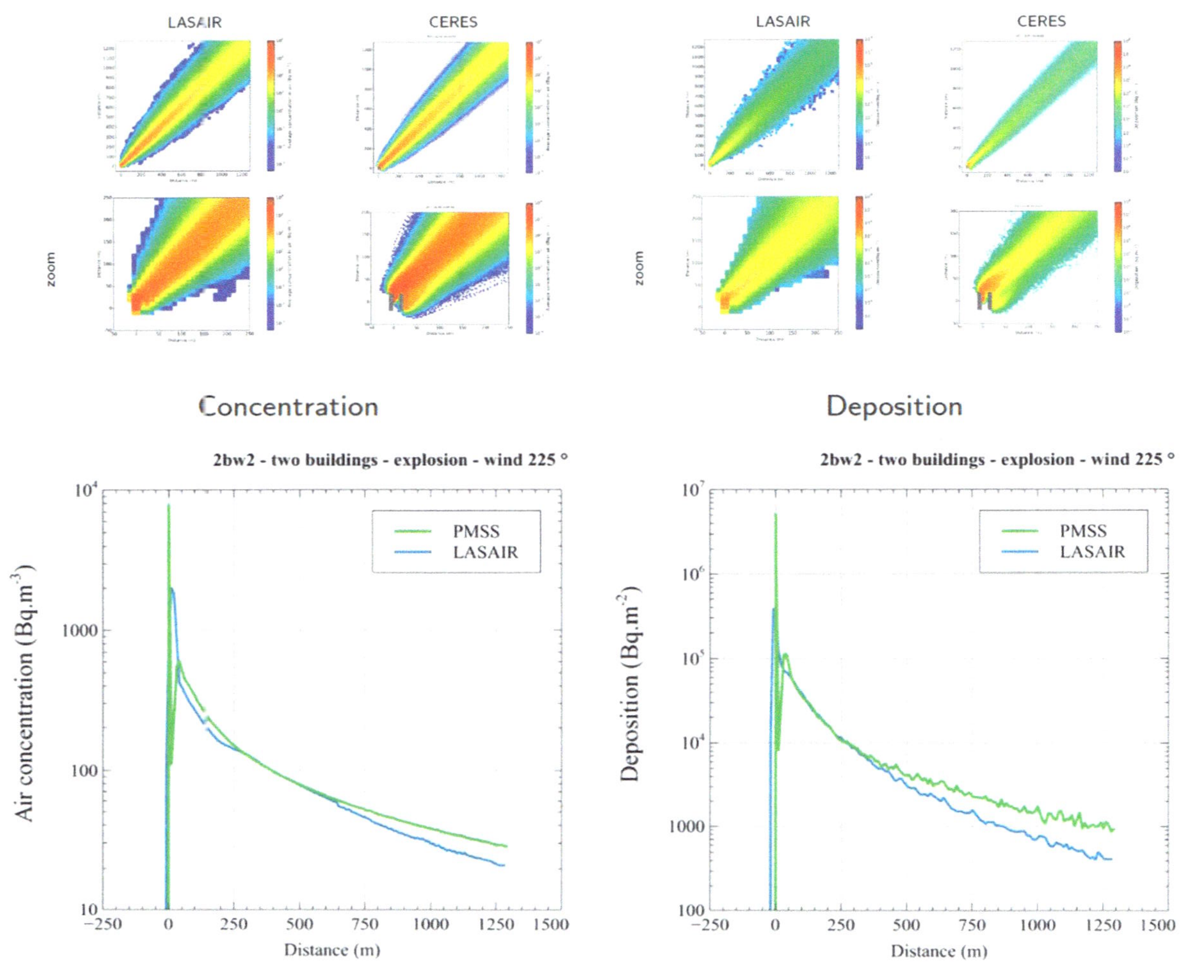

FIG. 13. Modelling results for Scenario 2bw2 (two buildings, explosion, wind 225°). The graphs on the left show predicted activity concentrations in air (Bq/m^3, averaged over 1 hour); the graphs on the right show predicted deposition after 1 hour (Bq/m^2). The first row of contour plots shows the predictions out to 1300 m downwind of the release point; the second row ('zoom') shows the predictions over the first 250 m downwind. The bottom graphs compare the predicted activity concentrations in air (left) or predicted deposition (right), as a function of downwind distance along the centre of the plume. (Both 'CERES' and 'PMSS' refer to predictions from CERES®CBRN-E.)

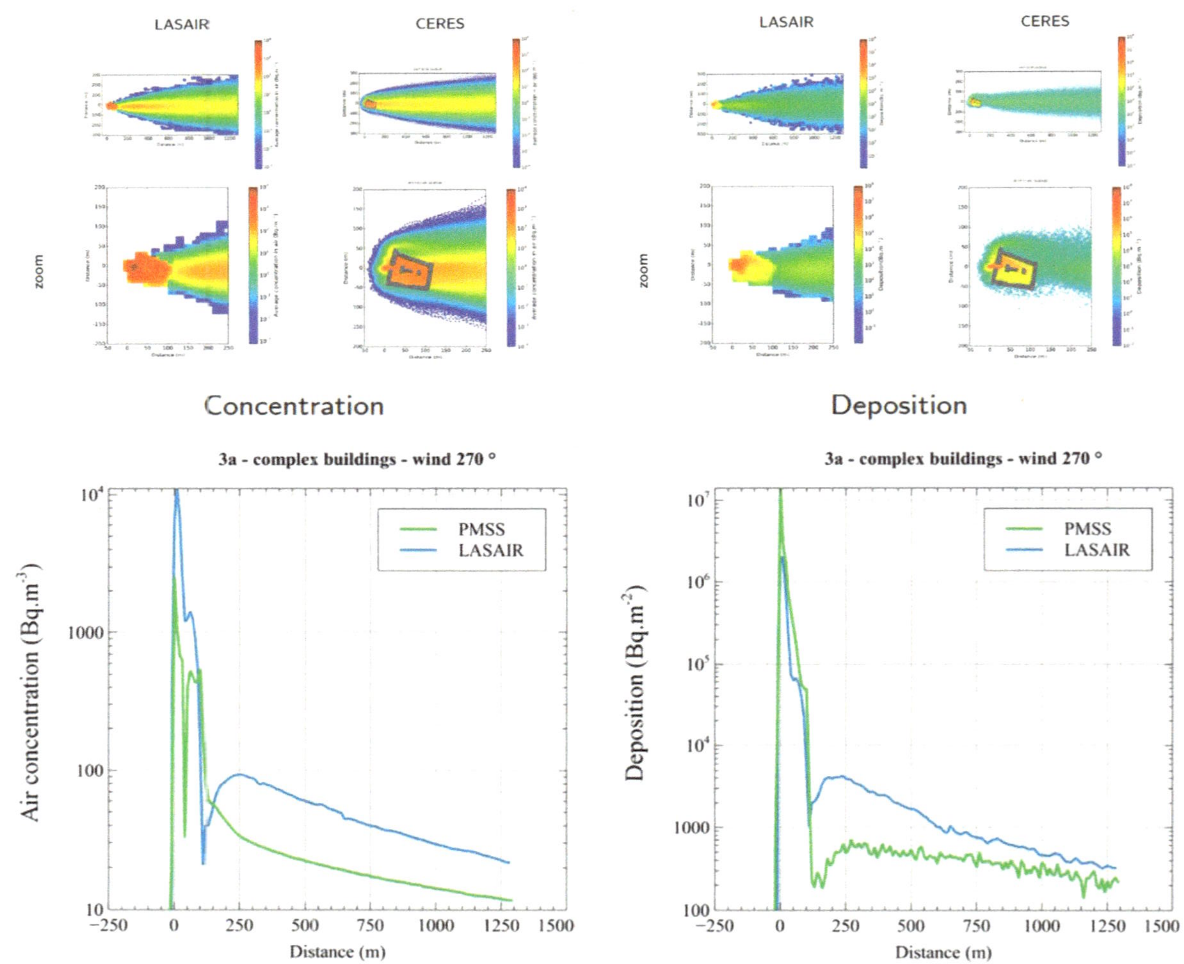

FIG. 14. Modelling results for Scenario 3a (complex buildings, continuous release, wind 270°). The graphs on the left show predicted activity concentrations in air (Bq/m^3, averaged over 1 hour); the graphs on the right show predicted deposition after 1 hour (Bq/m^2). The first row of contour plots shows the predictions out to 1300 m downwind of the release point; the second row ('zoom') shows the predictions over the first 250 m downwind. The bottom graphs compare the predicted activity concentrations in air (left) or predicted deposition (right), as a function of downwind distance along the centre of the plume. (Both 'CERES' and 'PMSS' refer to predictions from CERES®CBRN-E.)

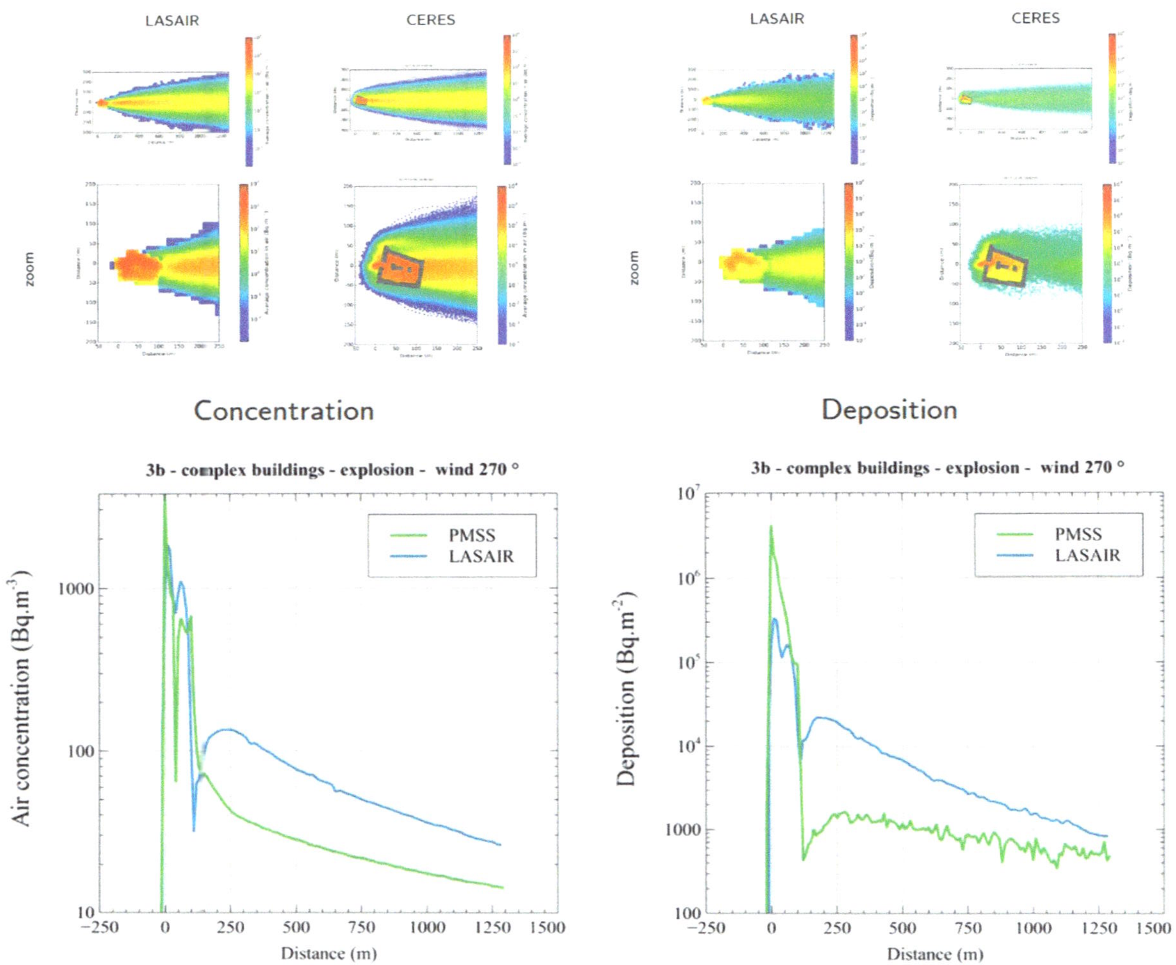

FIG. 15. Modelling results for Scenario 3b (complex buildings, explosion, wind 270°). The graphs on the left show predicted activity concentrations in air (Bq/m^3, averaged over 1 hour); the graphs on the right show predicted deposition after 1 hour (Bq/m^2). The first row of contour plots shows the predictions out to 1300 m downwind of the release point; the second row ('zoom') shows the predictions over the first 250 m downwind. The bottom graphs compare the predicted activity concentrations in air (left) or predicted deposition (right), as a function of downwind distance along the centre of the plume. (Both 'CERES' and 'PMSS' refer to predictions from CERES®CBRN-E.)

FIG. 16. Model predictions for Scenario 3 obtained with LASAIR for a 1 min continuous release, using different release points (-100 m, -50 m, -10 m, +60 m, +80 m). For the last column, the predictions assumed flat terrain (no buildings). The first row shows the dispersion point with respect to the building; the second row shows the predicted deposition after 1 h (Bq/m²). (Map data from OpenStreetMap https://www.openstreetmap.org/copyright, as used in the model LASAIR.)

3.5. FINDINGS FROM THE HYPOTHETICAL MUNICH EXERCISE

This exercise showed the effects of different building arrangements on activity concentrations in air and deposition in the vicinity of the buildings and in downwind areas, for a set of relatively simple conditions and two types of release. The effects of the buildings on the dispersion can be seen in the modelling results. For wind coming at an angle to the buildings (225°), the effects of the buildings appear to be less than for wind coming directly perpendicular to the buildings (270°). Building effects appear to be less at longer distances downwind than in the vicinity of the buildings, but this could be different with higher wind speeds and stable meteorological conditions; further exercises could examine a wider set of meteorological conditions.

The best agreement between the two models was obtained with Scenario 2, with the wind at 225°, and for which the building effects appeared to be less. CERES®CBRN-E appeared to give better precision and more realistic predictions in the vicinity of the obstacles. The types of turbulence scheme used in the two models, including the density assumed for the aerosols, could play a role in the discrepancies between the models, especially near the buildings.

In many of the figures, the activity concentrations in air and the depositions appear to be slightly higher for the explosion than for the continuous release. This is consistent with a higher effective release point for the explosion (cloud height, 16 m; Table 8) than for the continuous release (1 m; Table 8), compared with a building height of 20 m (Fig. 6).

4. MID-RANGE ATMOSPHERIC DISPERSION EXERCISE–ŠOŠTANJ

4.1. INTRODUCTION

During the MODARIA I programme, a scenario for use in mid-range atmospheric dispersion exercises was developed, based on measurement data obtained at the Šoštanj Thermal Power Plant (TPPŠ) in Slovenia [11, 13]. A set of measurements of SO_2 emissions and ambient monitoring data, together with detailed meteorological data, was obtained during a 3 week measuring campaign in 1991 [19]. These data provide a tracer experiment for the dispersion of an airborne contaminant from point sources (the stacks or chimneys of the power plant) to the surrounding area, over a complex terrain.

As previously described [11, 13], meteorological data were obtained in half-hour intervals at several ground-based weather stations (compliant with WMO standards) and with SODAR (vertical wind profiler). SO_2 emissions were measured automatically in half-hour intervals directly in the chimneys of the thermal power plant, and SO_2 concentrations in the region were measured automatically in half-hour intervals at measuring stations, positioned at key locations in the area. SO_2 concentrations from the power plant were very high in 1991, such that there was minimal measurement error and no confounding from other, much smaller, SO_2 sources in the region.

Two meteorological situations were modelled during the MODARIA I programme [11, 13]: (1) a simple situation with a strong wind blowing directly from a single operating chimney toward one measuring station atop a hill in the vicinity of the power plant; and (2) a complex situation with a nighttime temperature inversion followed by convective mixing, such that pollution occurred in several directions from the source. During MODARIA II, a third, intermediate, meteorological situation was modelled, as described in the rest of this section.

4.2. DESCRIPTION OF THE EXERCISE

This modelling exercise is based on a set of meteorological data, measurements of SO_2 emissions, and downwind pollution monitoring data obtained at the TPPŠ between 15 March 1991 and 5 April 1991. The test site, terrain, emissions sources, monitoring stations, meteorological equipment, and resulting data have been previously described in detail [11, 13]. The present exercise was based on a third meteorological situation of intermediate complexity between the two situations modelled during the MODARIA I programme, corresponding to the time period between 10:00 and 17:00 on 20 March 1991. High concentrations of air pollution were detected at three of the six monitoring stations (Fig. 17). Two chimneys were emitting SO_2 during that time period, with air pollution being spread in two main directions, i.e. first towards the west or northwest and later towards the east or northeast (Fig. 18).

Participants were provided with available data on meteorology and emissions, in addition to a digital elevation model for the area, data on corine land cover and surface roughness length, and basic environmental data (e.g. coordinates and altitudes of the emissions points and monitoring stations, and types of data recorded at each monitoring station). Data provided to participants included SODAR data (18 layers between 50 m and 1000 m height), meteorological data (temperature, relative humidity, air pressure, wind speed and direction, precipitation, global solar radiation), and emissions data (exhaust gas temperature, gas flow, and SO_2 concentration) at half-hour intervals for the time period. Participants were asked to predict the time dependent SO_2 concentrations for comparison with measured values at all six monitoring stations. All data were provided to participants in electronic format.

4.3. MODELS USED IN THE EXERCISE

Table 9 provides a summary of the two models used in this modelling exercise. More information about the individual models used in this exercise is provided in Appendix I of this publication and in Ref. [11].

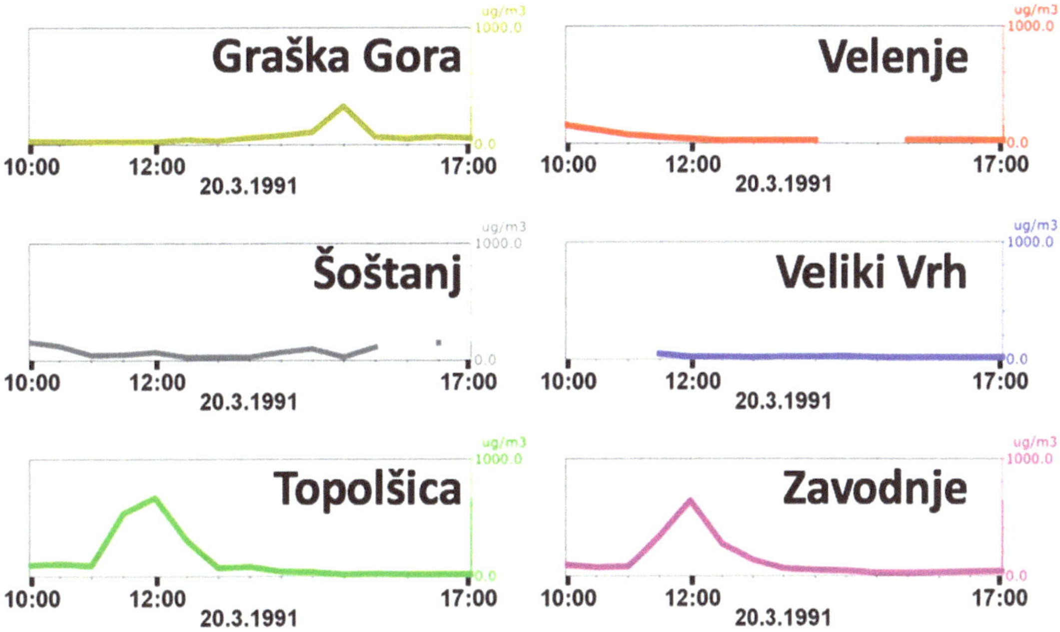

FIG. 17. Measured concentrations of air pollution (µg/m^3 SO_2) for 20 March 1991 (10:00–17:00) at the six monitoring stations.

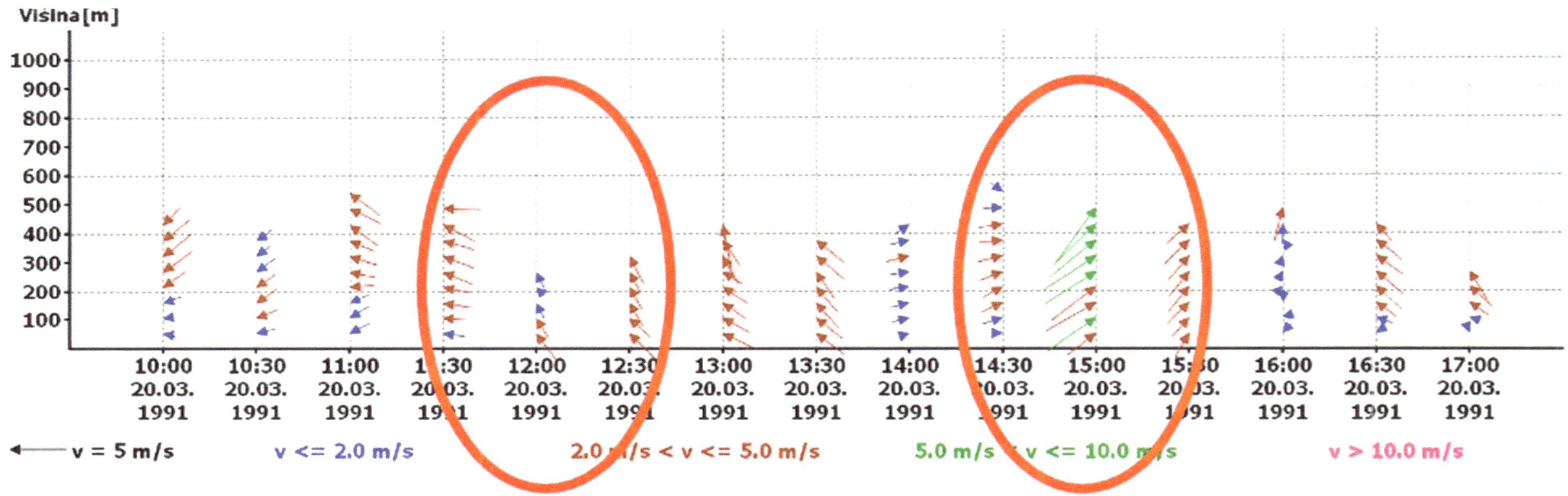

FIG. 18. SODAR data for 20 March 1991 (10:00–17:00). The direction of arrows represents the horizontal wind direction at the specified height. The length and colour of the arrow represent the horizontal wind speed at the specified height. Height is presented in meters above the ground (y-axis).

TABLE 9. SUMMARY OF MODELS AND MODELLING APPROACHES USED FOR THE ŠOŠTANJ EXERCISE

Model name	**SPRAY**	**ARTM v. 2.8**[a]
Participant and country	MEIS, Slovenia	M. Pattantyús-Ábrahám, Germany
Purpose of model	Fast dispersion microscale model for emergency response from nuclear power plant	Simulation of long term atmospheric dispersion from planned releases for regulatory purposes
Type of model	Lagrangian Particle Dispersion Model	Diagnostic wind field model + Lagrangian particle dispersion model
Number of Lagrangian particles	up to 2 000 000	1.008×10^9 per hour
Processes included	3D meteorological preprocessors SWIFT/MINERVE and SURFPRO	
Domain size/calculation range	15 km × 15 km × 6 km	User defined
Grid size	150 m × 150 m	100 m × 100 m
Grid height	Terrain following vertical coordinates, 20 layers up to 6 km, ground layer 12 m	User defined
Release height	100 m (B-1-2-3) 150 m (B-4) 230 m (B-5)	158 m
Receptor height	Ground level layer	Ground level
Stability class	Monin-Obukhov's length L calculated from meteorological measurements	Derived from wind sigma data of SODAR measurements at 100 m height; Monin-Obukhov length
Wind speed and direction	Measurements from 6 automatic measuring stations and SODAR, preprocessed by 3D wind field using MINERVE/SWIFT	Calculated wind and turbulence field by preprocessor TALdia
Air temperature	Measurements from 6 automatic measuring stations preprocessed by using SURFPRO	Hourly averages of measurements
Dispersion parameters	Calculated from measurements preprocessed by 3D turbulence preprocessor SURFPRO	Diagnostic wind fields based on SODAR data at emission height (158 m) and approximate height of neighbouring hills (376 m); hourly averages of measurements
Plume rise	Calculated from measurements preprocessed by 3D meteorological preprocessor	Calculated from heat emission, based on the temperature of the exhaust gas
Release time step	10 s	Hour
Calculation time step	1800 s	Hour
Simulation time	ca. 33 hours for complete duration of campaign (from 15.03.1991 to 05.04.1991)	
Inversion layer height	Calculated from measurements preprocessed by 3D meteorological preprocessor	
Terrain/topography	Complex, Corine land use cover used (21 categories)	Detailed orography of nonhomogeneous terrain
Friction coefficient or friction velocity	Included in SURFPRO, calculated based on the real measured data and 2D geographical data	
Rugosity	Included in SURFPRO, calculated based 2D geographical data	Manually determined
Time to set up and run	At least 2 weeks	
Time to process results	At least 1 week	
Total source term	ca. 500 t of SO_2 from B-1-2-3 ca. 1000 t of SO_2 from B-4 ca. 1500 t of SO_2 from B-5	
Reference(s) for code (URL)	http://www.aria.fr	Ref. [20]

4.4. ANALYSIS OF MODELLING RESULTS

4.4.1. SPRAY model

Modelling was carried out with the operational version of the SPRAY model. Modelling results compared with measurements for each monitoring station are shown in Fig. 19. Contour maps of the modelled concentrations of SO_2 at two time points (13:00 and 16:00) are shown in Fig. 20.

Figures 21–26 show the results of the operational version of the model (using traditional validation methodology) and those using an 'enhanced validation methodology' [11]. The upper left graphs show the model predictions for the specified location and time (operational version). The upper right graphs show the model predictions for the specified location and three cells in each direction, corresponding to the best fit to the measurements within the range of location and time. The lower graphs show the model predictions for the specified time period and the preceding and following time periods ($3\Delta t$), for the specified location and either three (left) or five (right) cells in each direction, corresponding to the best fit to the measurements within the range of location and time. This approach to allowing for uncertainty about position and time in the model results was described in detail in the MODARIA I report [11].

For the Graška Gora monitoring station (Fig. 21), the operational version of the model missed both the timing and the size of the peak concentration (15:00). Allowing for uncertainty in position and timing of the peak improved the model predictions with respect to timing of the peak, but the model predictions still underestimated the peak measurement. Similar small improvements were seen for the Šoštanj (Fig. 22), Velenje (Fig. 24), and Veliki Vrh (Fig. 25) stations, although the differences among model runs were small. For both the Topolšica and Zavodnje stations, the model was unable to reproduce the major peak concentrations, although the timing of the predicted peaks improved slightly when adjacent cells or time periods were considered.

Thus, in practice, this meteorological case turned out to be more complex than expected, suggesting that additional information would need to be considered in order to reproduce the observed peaks in SO_2 concentrations at three of the monitoring stations.

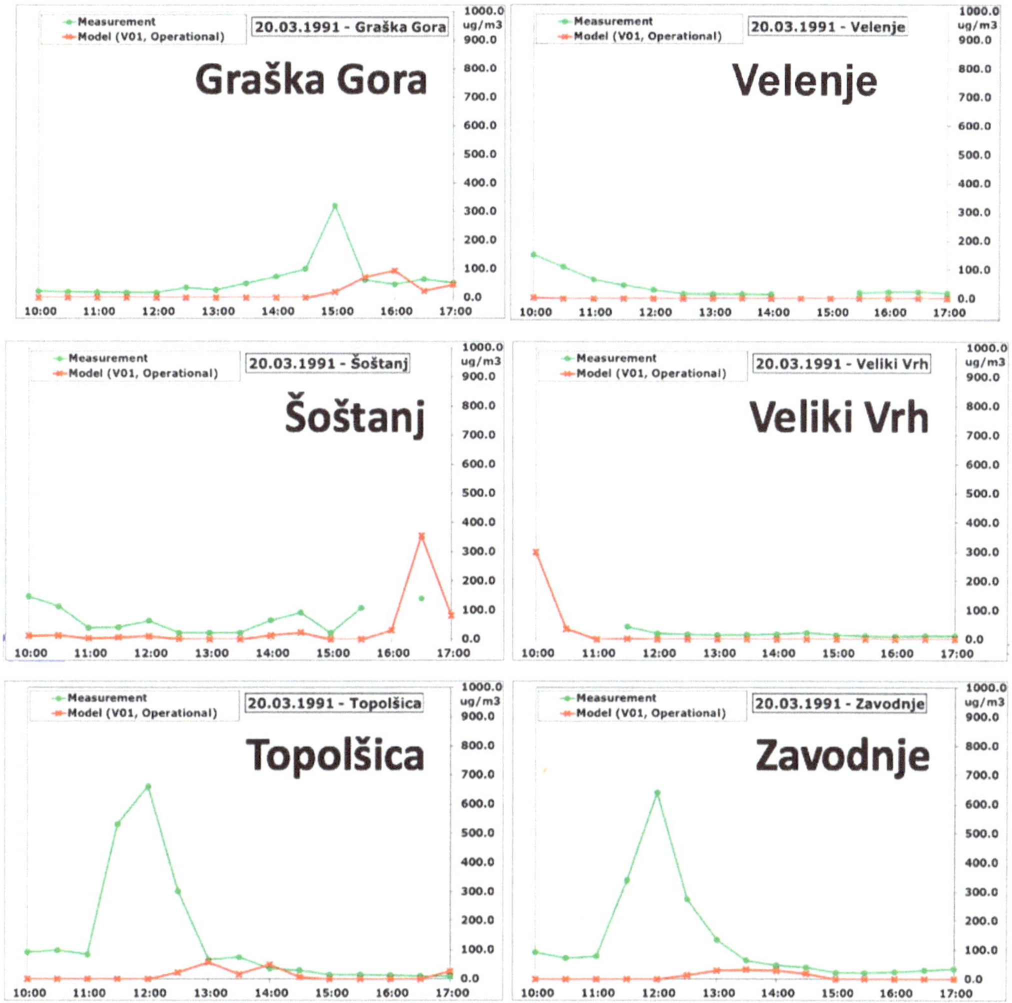

FIG. 19. Comparison of model predictions (red) using SPRAY with the measurements (green) of SO_2 ($\mu g/m^3$) for 20 March 1991 (10:00–17:00) at the six monitoring stations.

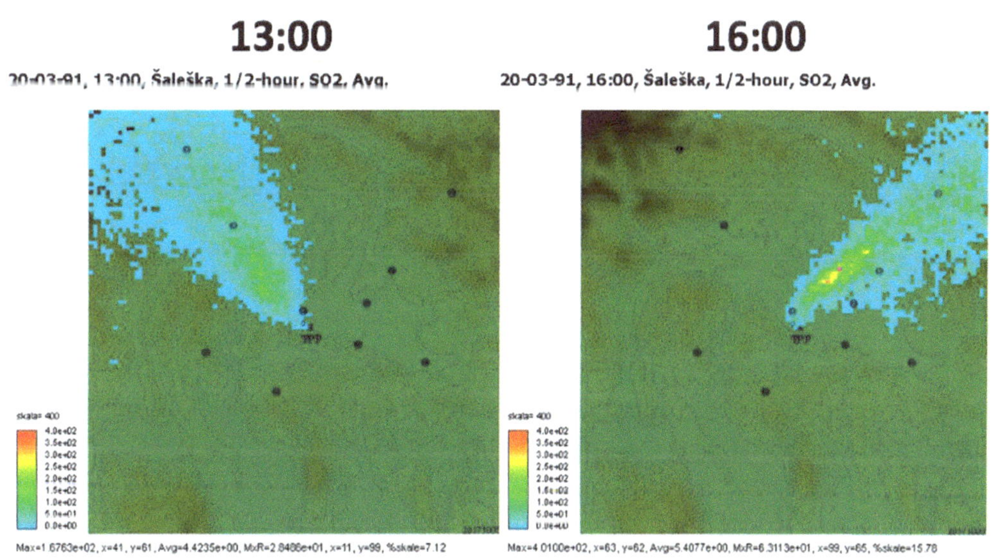

FIG. 20. Contour maps of predicted air pollution using the SPRAY model for two time points on 20 May 1991.

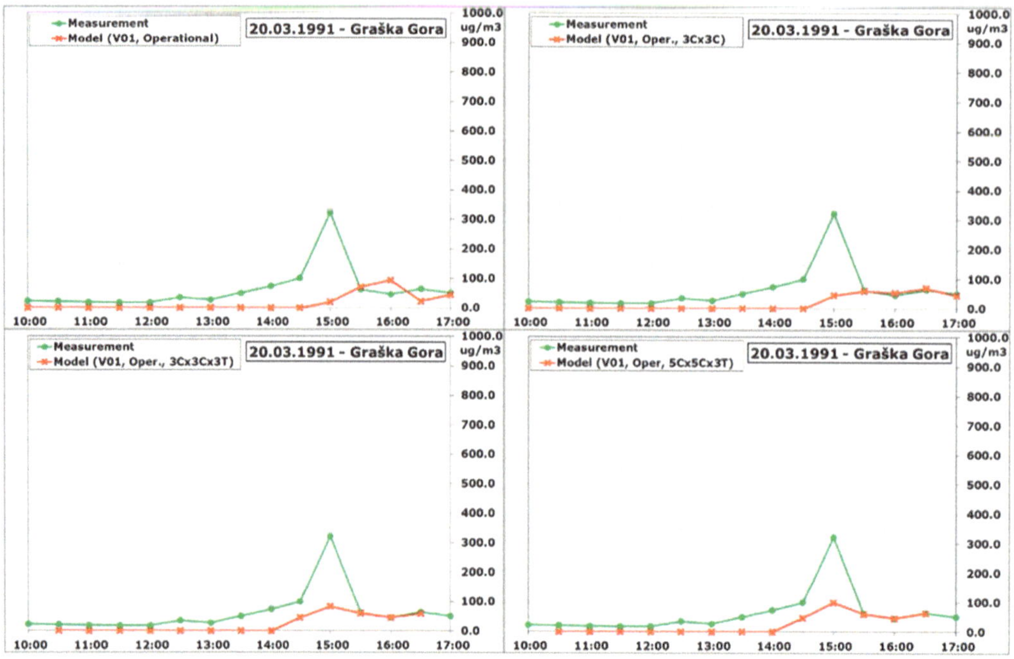

FIG. 21. Comparison of model predictions (red) using SPRAY with the measurements (green) of SO_2 ($\mu g/m^3$) for the period 20 March 1991 (10:00–17:00) at the Graška Gora station. The upper left graph shows results for the specified location and time obtained with the operational version. The remaining graphs show the best fit to the measurements for a range of location (3 or 5 cells in each direction) and (lower graphs) time periods ($\pm 3\Delta t$).

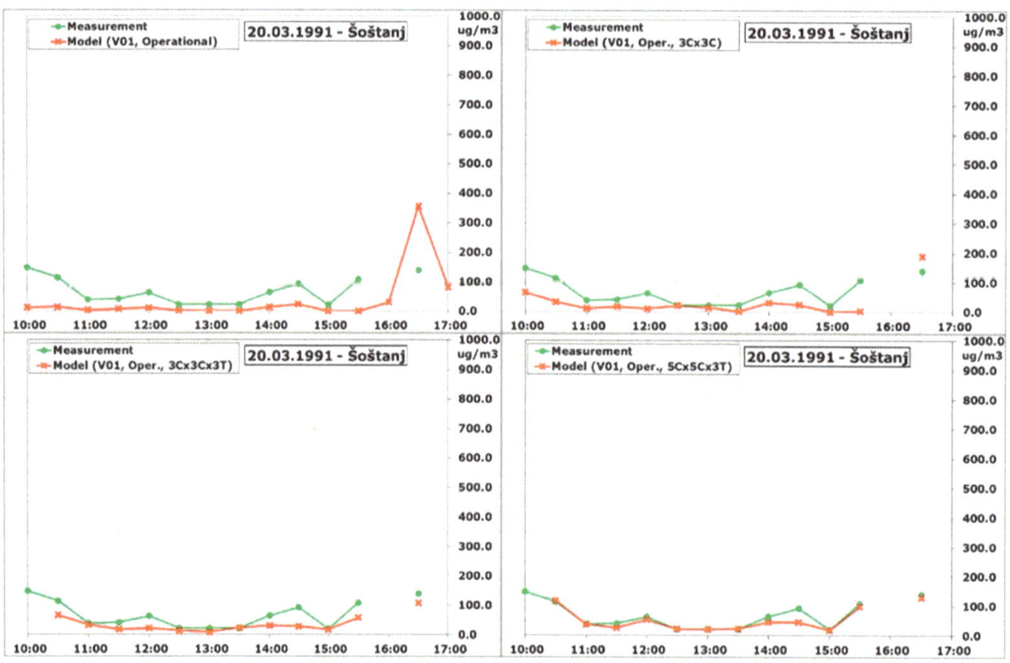

FIG. 22. Comparison of model predictions (red) using SPRAY with the measurements (green) of SO_2 ($\mu g/m^3$) for the period 20 March 1991 (10:00–17:00) at the Šoštanj station. The upper left graph shows results for the specified location and time obtained with the operational version. The remaining graphs show the best fit to the measurements for a range of location (3 or 5 cells in each direction) and (lower graphs) time periods ($\pm 3\Delta t$).

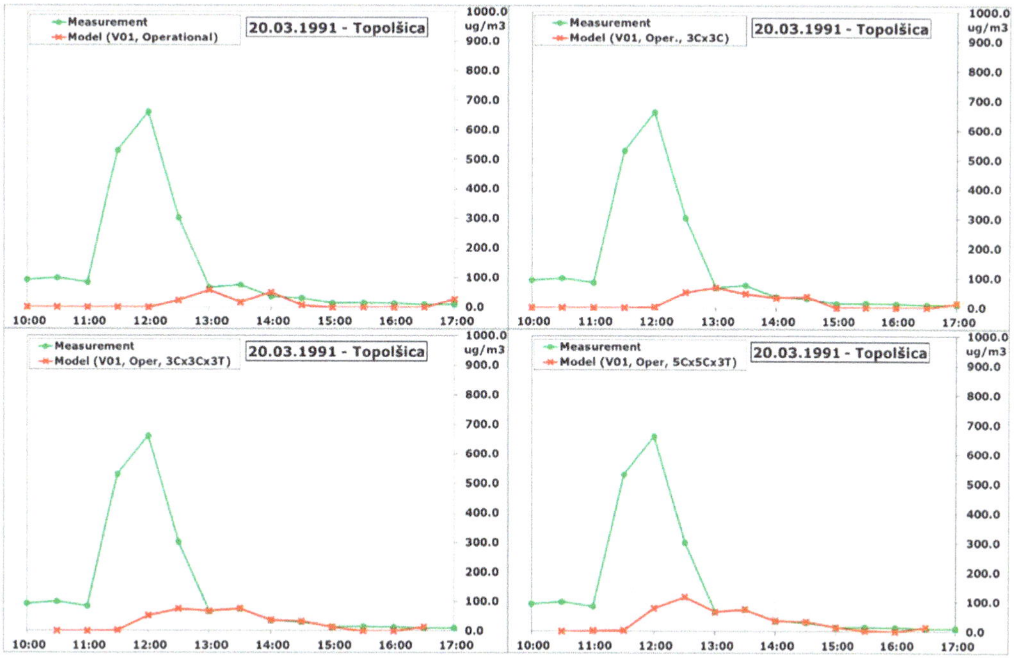

FIG. 23. Comparison of model predictions (red) using SPRAY with the measurements (green) of SO_2 ($\mu g/m^3$) for the period 20 March 1991 (10:00–17:00) at the Topolšica station. The upper left graph shows results for the specified location and time obtained with the operational version. The remaining graphs show the best fit to the measurements for a range of location (3 or 5 cells in each direction) and (lower graphs) time periods ($\pm 3\Delta t$).

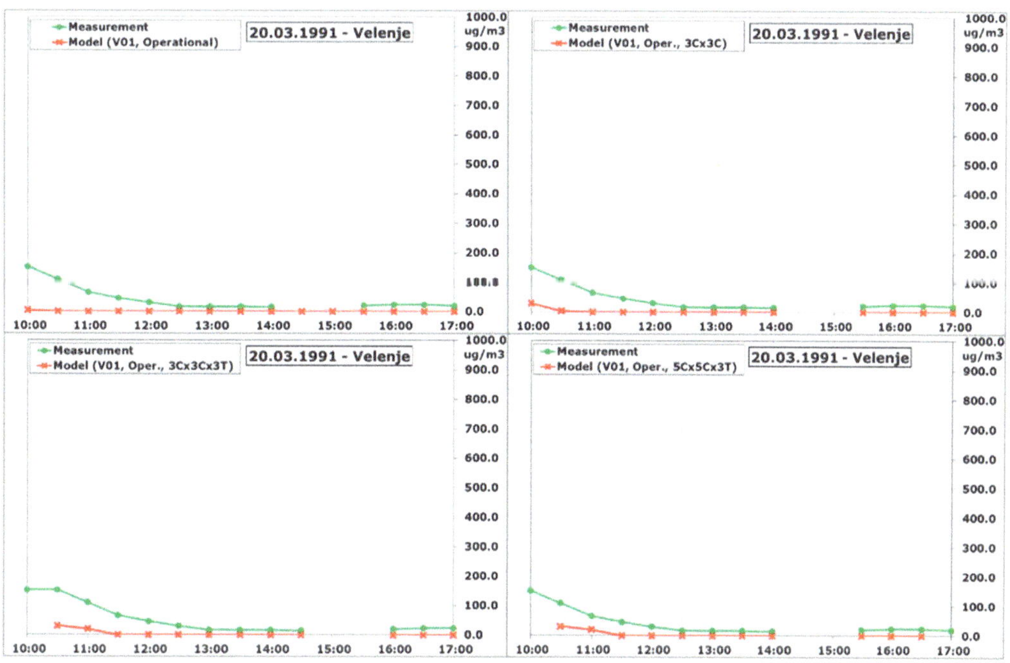

FIG. 24. Comparison of model predictions (red) using SPRAY with the measurements (green) of SO_2 ($\mu g/m^3$) for the period 20 March 1991 (10:00–17:00) at the Velenje station. The upper left graph shows results for the specified location and time obtained with the operational version. The remaining graphs show the best fit to the measurements for a range of location (3 or 5 cells in each direction) and (lower graphs) time periods ($\pm 3\Delta t$).

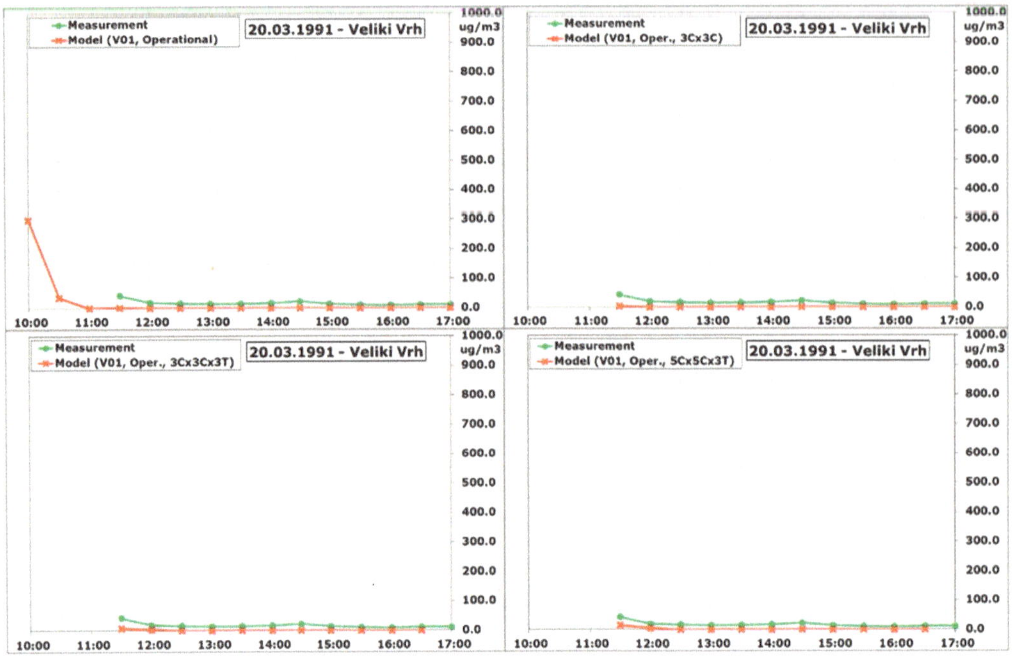

FIG. 25. Comparison of model predictions (red) using SPRAY with the measurements (green) of SO_2 ($\mu g/m^3$) for the period 20 March 1991 (10:00–17:00) at the Veliki Vrh station. The upper left graph shows results for the specified location and time obtained with the operational version. The remaining graphs show the best fit to the measurements for a range of location (3 or 5 cells in each direction) and (lower graphs) time periods ($\pm 3\Delta t$).

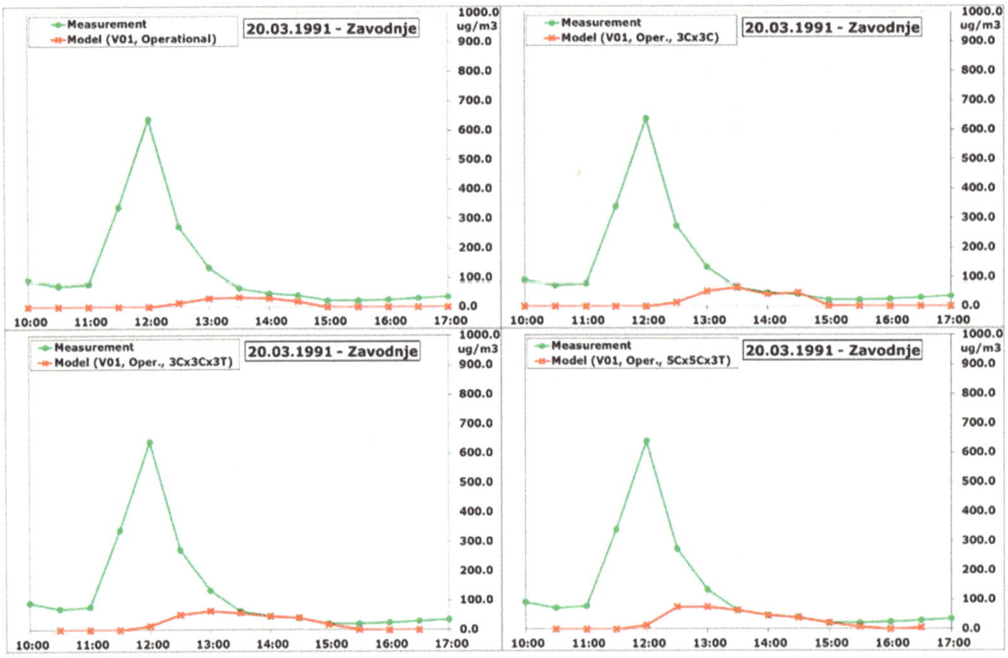

FIG. 26. Comparison of model predictions (red) using SPRAY with the measurements (green) of SO_2 ($\mu g/m^3$) for the period 20 March 1991 (10:00–17:00) at the Zavodnje station. The upper left graph shows results for the specified location and time obtained with the operational version. The remaining graphs show the best fit to the measurements for a range of location (3 or 5 cells in each direction) and (lower graphs) time periods ($\pm 3\Delta t$).

4.4.2. ARTM model

Modelling was carried out with the ARTM v. 2.8 model [20], using SODAR wind data at two different heights, one corresponding to the emission height (158 m) and the other corresponding to the approximate height of the neighbouring hills (376 m). Other parameters were the same for both simulations. Orography (terrain) and weather dependent plume rise were accounted for. Details of the model and the parameterisation are summarized in Appendix I. Modelling results compared with measurements for each monitoring station are shown in Fig. 27. The measurements are shown both in half-hour intervals as reported and as hourly averages, because ARTM delivers only hourly outputs.

As seen in Fig. 27, the reported (half-hourly) and hourly average measurements coincided when SO_2 concentrations were low (Velenje, Veliki Vrh, non-peak times for the other four stations). However, where peaks occurred, generally the hourly averages exceeded the half-hourly measurements. The model predictions overestimated the peak concentration by differing amounts (Graška Gora), predicted peaks where the measurements did not display much of a peak (Šoštanj), or misplaced the peak with respect to time (Zavodnje; 376 m prediction for Topolšica). The predictions for Topolšica using the 158 m wind data were quite close to the half-hourly measurements. Predictions with the 158 m and 376 m wind data were very close to each other for Graška Gora, Šoštanj (first predicted peak), and Zavodnje, while for Šoštanj (second predicted peak) and Topolšica, the peak obtained with the 158 m wind data exceeded that obtained with the 376 m wind data.

Figure 28 shows contour maps of the model predictions using the two different sets of wind data. For both the 158 m and 376 m wind data, the predicted contamination spread in all directions except southeast. For the 158 m predictions, the maximum predicted SO_2 concentration was about 4 km to the west of the source. For the 376 m predictions, the maximum predicted SO_2 concentration was about 3500 m to the north of the source, with considerably higher predicted contamination to the northwest than was obtained with the 158 m wind data. Results with the two sets of wind data were otherwise generally similar.

4.5. FINDINGS FROM THE ŠOŠTANJ EXERCISE

This exercise turned out to be more complex to model than had been anticipated. Both models predicted plumes generally toward the northwest and northeast, but the predicted SO_2 concentrations were generally lower with SPRAY than with ARTM, and (with the ARTM model) lower when calculated using the 158 m wind data than the 376 m wind data. Peak concentrations at specific locations were difficult to model, possibly reflecting difficulties in sufficiently characterizing the windfield above the terrain. Allowing for spatial and temporal error improved the model predictions with SPRAY.

4.6. ACKNOWLEDGEMENT

The projects, L2-8174 Method for the forecasting of local radiological pollution of atmosphere using Gaussian process models, and L2-2615 Modelling the Dynamics of Short-Term Exposure to Radiation, were financially supported by the Slovenian Research Agency.

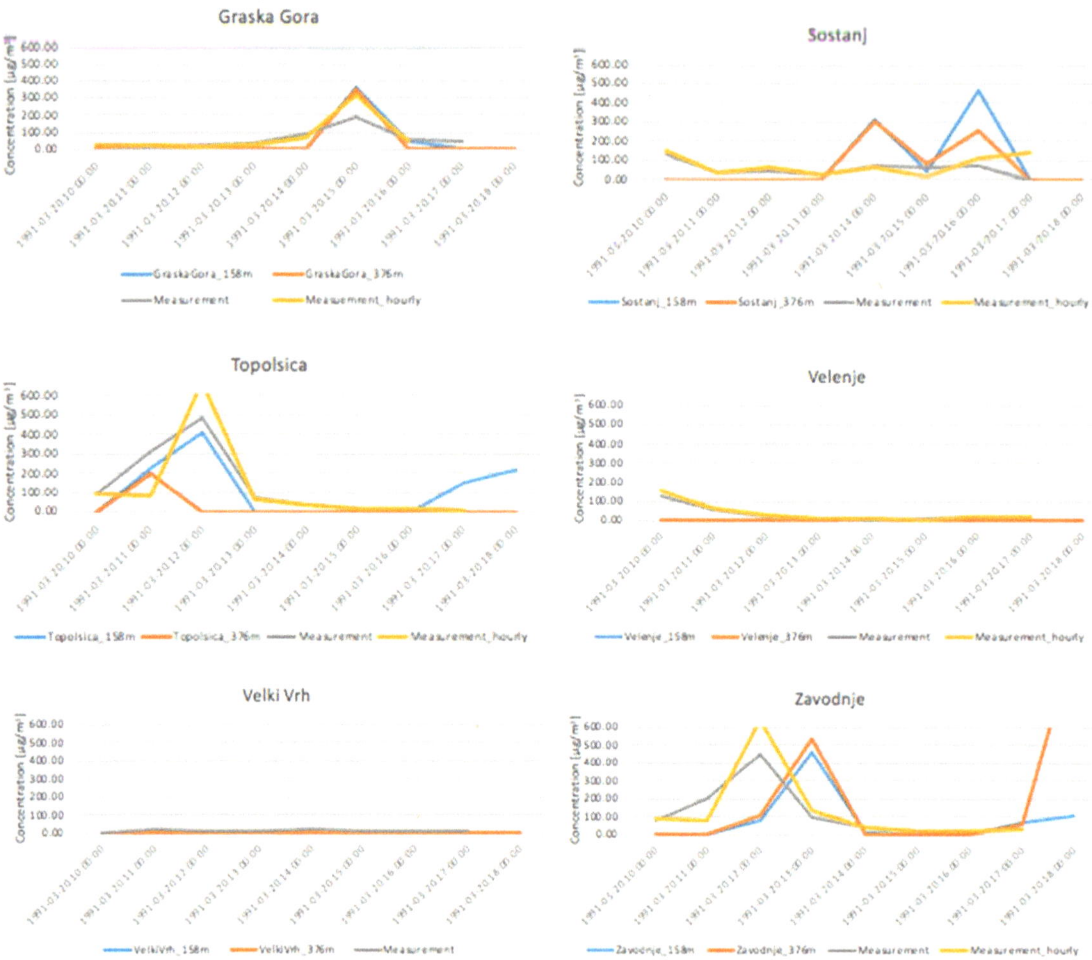

FIG. 27. Comparison of model predictions using ARTM with the measurements of SO_2 ($\mu g/m^3$) for 20 March 1991 (10:00–17:00) at the six monitoring stations. Model results are shown for the SODAR windfield at 158 m (blue) and 376 m (orange). Measurements are shown as reported on the hour (yellow) or as averages of two half-hour measurements (grey).

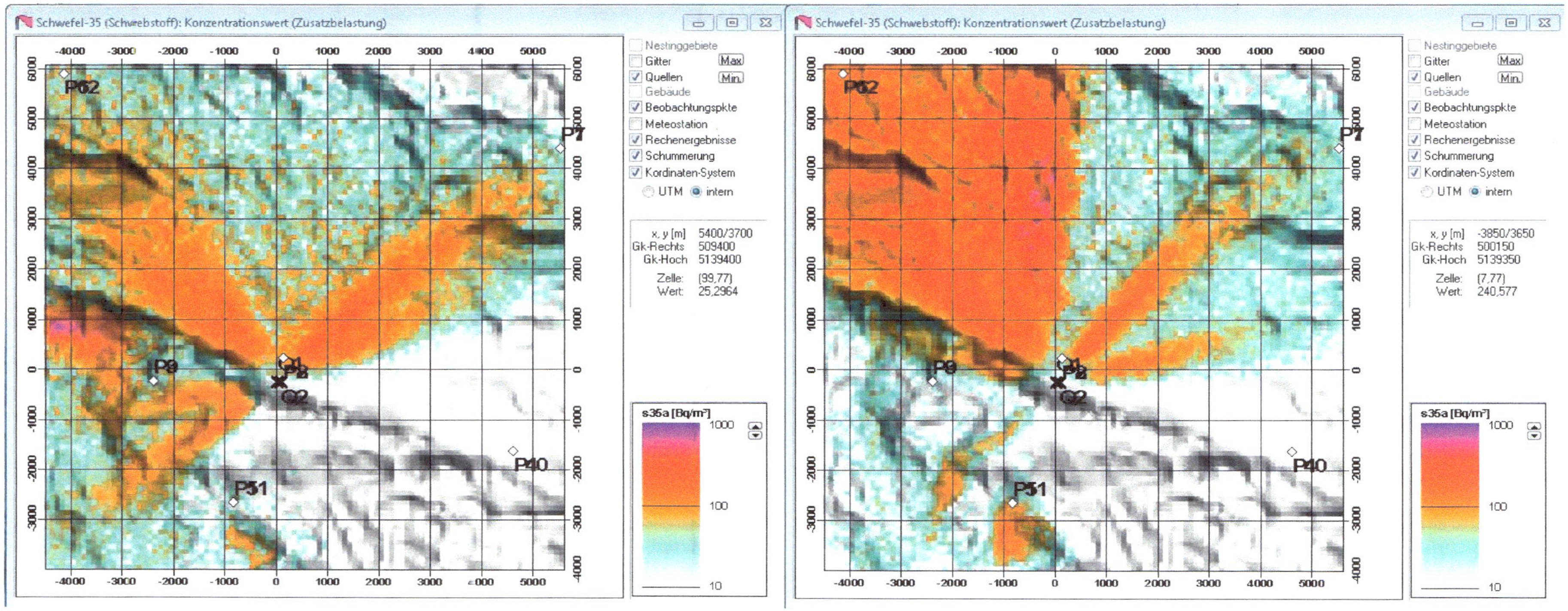

FIG. 28. Contour maps of predicted air pollution using the ARTM model (Bundesamt für Strahlenschutz, Germany) with SODAR windfields obtained for 158 m (left) and 376 m (right). Black Xs denote the locations of the stacks (Block 1, 2, 3 and Block 4). White diamonds indicate the locations of the monitoring stations.

5. WEATHER FORECASTING FOR MODELLING ATMOSPHERIC DISPERSION

5.1. INTRODUCTION

Atmospheric dispersion modelling of accidental releases of radioactivity (e.g. after an accident at a nuclear installation) is used to generate spatial maps of radionuclide concentrations in the vicinity of the release source. The modelled spatial distribution of radioactivity may then be used in decision making on the implementation of protective actions for the public. These atmospheric dispersion models depend on meteorological information in the vicinity of the source of airborne contamination.

Typically, the meteorological data serve as inputs to the atmospheric dispersion model and are handled in a preprocessor incorporated into the model. Two approaches may be used with meteorological data, a diagnostic approach or a prognostic approach. A diagnostic approach uses only measurements of meteorological parameters; the measurements may be handled in real time (as the measurements are obtained) or after the fact. A prognostic approach uses forecasts of meteorological parameters (weather forecasts) to predict the meteorological data and consequent atmospheric dispersion of any contamination. A prognostic approach allows planning ahead, e.g. before the start of planned emissions or to predict the likely movement of the plume from an accidental release. For some specific analyses, measurement data and prognostic meteorological information can be combined; typically measurements are used as available, with prognostic information used when measurements are not available, e.g. for higher levels of the atmosphere beyond the range of the measuring equipment.

The Šoštanj modelling exercise described in Section 4 of this publication and previously used in MODARIA I [11, 13] was based on a detailed set of meteorological data in Slovenia for a three-week period in 1991 [19, 21]. In addition to the atmospheric dispersion modelling exercises based on this data set, the Šoštanj meteorological data set provided an opportunity to compare diagnostic and prognostic approaches to handling meteorological data. The measurements made in 1991 serve as input for a diagnostic approach and forecasting data for the location and time period serve as input for a prognostic approach. This section describes efforts to assess the quality of meteorological forecasts as a basis for dispersion modelling (see also Ref. [15]).

5.2. METHODOLOGY

High quality data sets with measurements from real life situations are needed to assess the forecasting quality of atmospheric dispersion capability. Although there are data from two major nuclear accidents (Chernobyl and Fukushima), the applicable measured meteorological data for the immediate area around either facility are insufficient. In addition, the time development and quantity of emissions and the emission temperature and volume flows are also unknown. It is therefore not possible to make a reliable assessment of the modelling quality based on a data set from either of those accidents. In light of this, the data set 'Šoštanj 91' was chosen for this test; in this data set, the course of the dispersion was tracked via a chemical tracer. A brief description of the characteristics of the experiment are presented, with further details available from Refs [19, 21].

5.2.1. The 'Šoštanj 91' experiment

The data set used for this exercise has been described previously in Section 4 above (see also Refs [11, 13]). Detailed data are available for 15 March to 7 April 1991. Measurements are available for 30 minute increments.

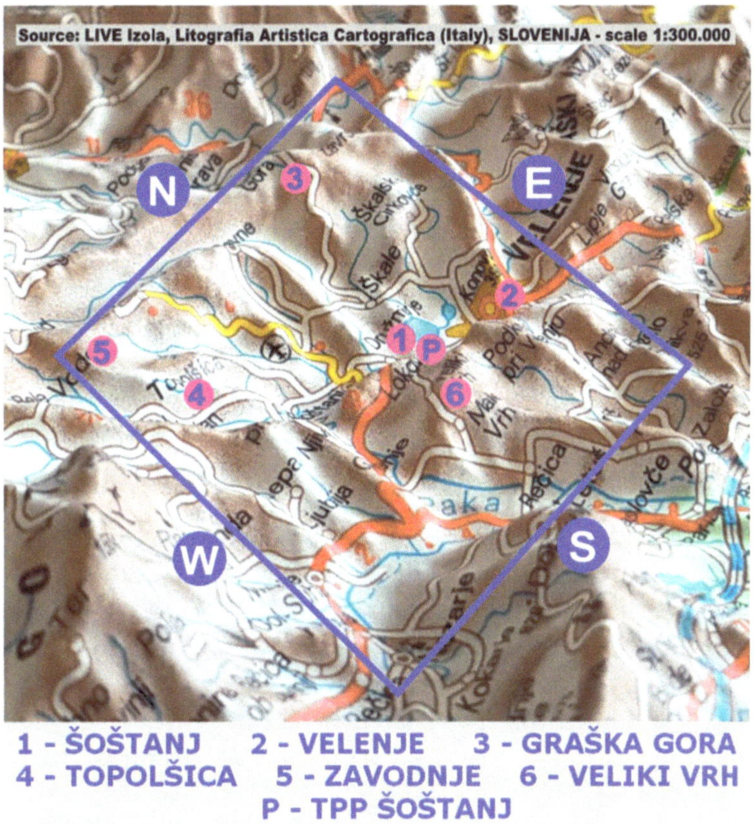

FIG. 29. Target modelling area and distribution of meteorological stations. (Source: LIVE Izola, Litografia Artistica Cartografica, Italy.) reproduced with permission courtesy of Harmo Initiative.

Weather models were tested retroactively for the time period of the measurements. Initial and boundary conditions were taken from re-analyses (GFS; see Section 5.2.4). This method was used primarily to test weather models and to reduce the impact of possible errors in initial and boundary conditions[3].

5.2.2. Test site

Validation of the meteorological forecasting was performed for a 15 km × 15 km area around the Šoštanj facility (Fig. 29), corresponding to the area for which detailed measurements are available for key meteorological variables. Modelling was performed for several domains with different resolutions (nested domains).

5.2.3. Sources of forecasting data

Several sources of weather forecasting data are available, as follows:

— The NCEP Climate Forecast System (CFSR) developed by the National Centers for Environmental Prediction (NCEP) of the National Oceanic and Atmospheric Administration (NOAA), National Weather Service, has published The NCEP Climate Forecast System Reanalysis (1979–2010) and The NCEP Climate Forecast System

[3] Initial conditions are for the time period being modelled. Boundary conditions have to do with the meteorological situations at the edges of the geographical or atmospheric regions being modelled.

Version 2 [22, 23] (see also: http://cfs.ncep.noaa.gov/). The CFSR provides worldwide meteorological information from 1979 to 2010, at a resolution of 0.5° and 6 hours;

- The Global Forecast System (GFS) is a global numerical weather prediction system from NCEP that generates data for a variety of atmospheric and land–soil variables (https://www.ncei.noaa.gov/products/weather-climate-models/global-forecast). The GFS produces forecasts four times daily for up to 16 days in advance (https://www.emc.ncep.noaa.gov/emc/pages/numerical_forecast_systems/gfs.php);
- The European Centre for Medium-Range Weather Forecasts (ECMWF; http://apps.ecmwf.int/datasets/) provides several global analyses (ERA, for ECMWF Re-Analysis), including ERA-40 (September 1957–August 2002) and ERA-Interim (January 1979–August 2019). ERA-Interim has a resolution of 0.7° and 6 hours.

5.2.4. Forecasting model

The Weather Research and Forecasting (WRF) model (most commonly WRF-ARW, Advanced Research WRF; https://www.mmm.ucar.edu/weather-research-and-forecasting-model) is a mesoscale numerical weather prediction system designed for both atmospheric research and operational forecasting applications. WRF-ARW is commonly used with data sets such as CFSR or ERA-Interim to produce prognostic analyses of meteorology for specified locations and dates. Typically, the forecasting is handled in terms of nested models with increasing resolution with respect to geography (e.g. from a cell size of several km to a cell size of 1 km).

The WRF-ARW model was selected for this exercise. This model is designed to have a fine resolution, both spatial and temporal, in order to handle the meteorology for an area with complex terrain. As described in Section 4 (see also Refs [11, 13]) the terrain in the vicinity of the Šoštanj facility is extremely complex; therefore, the modelling was done with a higher resolution (smaller grid) than had previously been validated for the WRF model, and thus the present work is an advancement from previous efforts.

5.2.5. Testing – four simulations

Four simulations to test weather forecasts were performed, the key characteristics for which are summarized in Table 10. These four similar, but not identical, simulations were carried out to establish to what extent different model setups affect the quality of the end results.

5.3. RESULTS

The results are described below in terms of model predictions for four key weather variables that drive the atmospheric dispersion modelling. These variables include air temperature, wind speed (measurements at several locations), global solar radiation and precipitation (measurements at one site). Air temperature and wind speed were measured at several locations, while global solar radiation and precipitation were measured at one location.

Both measurements and forecasts were analysed in 30-minute intervals. The main interests were prediction of the timeline for each variable, matching of the 30-minute intervals, statistical matching of daily cycles (including the distribution of daily errors), and matching of numerical statistical estimators between the measured and forecast values. For statistical estimates, the following coefficients were used: normalized mean square error (NMSE), fractional bias (FB), mean square error (MSE), root mean square error (RMSE), Pearson correlation coefficient (R) and coefficient of determination (R^2) [24, 25]. Sunflower plots were used for analysis of daily cycles and daily error distributions [26, 27].

TABLE 10. KEY CHARACTERISTICS OF THE FOUR SIMULATIONS[a]

Group/ Simulation no.	Weather model	Inputs and boundary conditions	Resolution	Domain 1 spatial and temporal resolution, grid	Domain 2 spatial and temporal resolution, grid	Domain 3 spatial and temporal resolution, grid
MEIS-1 (GFS-MEIS)	WRF ARW 3.9.1	NCEP climate forecast system reanalysis (CFSR)[b]	0.5° 6 h	25 km 3 h 80 × 80	5 km 0.5 h 86 × 86	1 km 0.5 h 101 × 101
MEIS-2 (ERA-MEIS)	WRF ARW 3.9.1	ERA Interim (ECMWF)[c]	0.7° 6 h	25 km 3 h 80 × 80	5 km 0.5 h 86 × 86	1 km 0.5 h 101 × 101
CEA-1 (GFS-CEA)	WRF ARW 4.0	NCEP climate forecast system reanalysis (CFSR)[b]	0.5° 6 h	25 km 3 h 80 × 80	5 km 1 h 86 × 86	1 km 0.5 h 101 × 101
CEA-2 (ERA-CEA)	WRF ARW 4.0	ERA Interim (ECMWF)[c]	0.7° 6 h	25 km 3 h 80 × 80	5 km 1 h 86 × 86	1 km 0.5 h 101 × 101

[a] Table adapted from Ref. [15], used with permission courtesy of Harmo Initiative.
[b] See Ref. [22].
[c] See Ref. [28].
Abbreviations:
CEA: Commissariat à l'Energie Atomique, France
CFSR: Climate Forecast System Reanalysis (1979–2010) from the National Oceanic and Atmospheric Administration, National Weather Service, National Centers for Environmental Prediction (https://cfs.ncep.noaa.gov)
ECMWF: European Centre for Medium-Range Weather Forecasts (http://apps.ecmwf.int/datasets/)
ERA: ECMWF Re-Analysis (several versions of ERA are available)
GFS: Global Forecast System (https://www.ncei.noaa.gov/products/weather-climate-models/global-forecast)
MEIS: MEIS d.o.o., Slovenia
WRF: Weather Research and Forecasting (https://www.mmm.ucar.edu/weather-research-and-forecasting-model)
WRF – ARW: Weather Research and Forecasting – Advanced Research WRF

5.3.1. Air temperature

To provide the reader with a reference point regarding the main air temperature characteristics of the period in the spring of 1991, an air temperature profile chart was simulated for each major source of forecasting data (Figs 30 and 31). Unfortunately, it was not possible to validate the temperature profile as a whole in the absence of data from radio acoustic sounding systems (RASS). However, it was possible to validate individual ground level values at the sites of the meteorological measurement stations (Figs 32–42).

Figure 32 provides a statistical analysis of the predicted and measured temperatures at each of five meteorological stations for the period 15 March 1991 to 5 April 1991. Figures 33 and 34 show a comparison of the measured and modelled temperature over time at each of the five meteorological stations, for each of the four simulations. Figures 35–38 provide scatter plots for the measured and modelled temperature data at the five meteorological stations, for each of the four simulations, together with several statistical parameters (R, R^2, RMSE, and MSE). Figures 39–42 illustrate the errors (modelled temperature – measured temperature) by time of day (daily cycle) for each of the five meteorological stations and each of the four simulations.

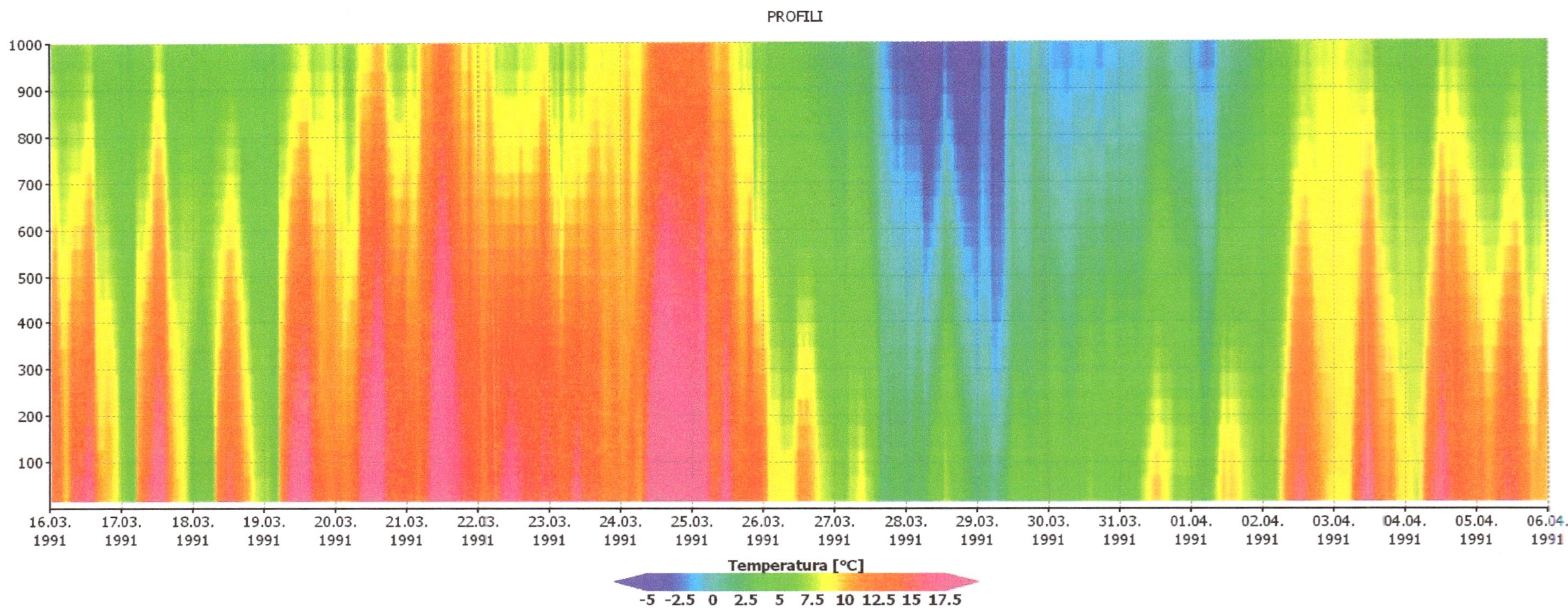

FIG. 30. Vertical air temperature profile, simulation results from MEIS-1.

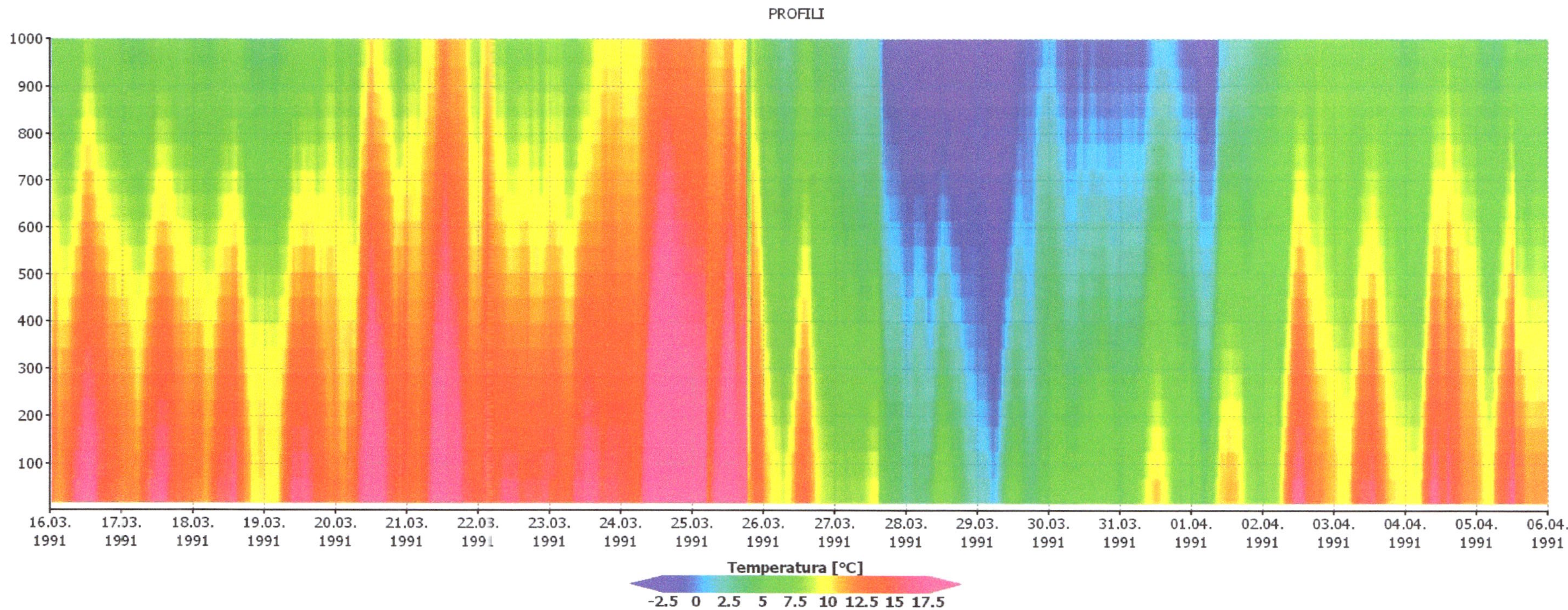

FIG. 31. Vertical air temperature profile, simulction results from MEIS-2 (reproduced from Ref.[15], with permission courtesy of Harmo Initiative).

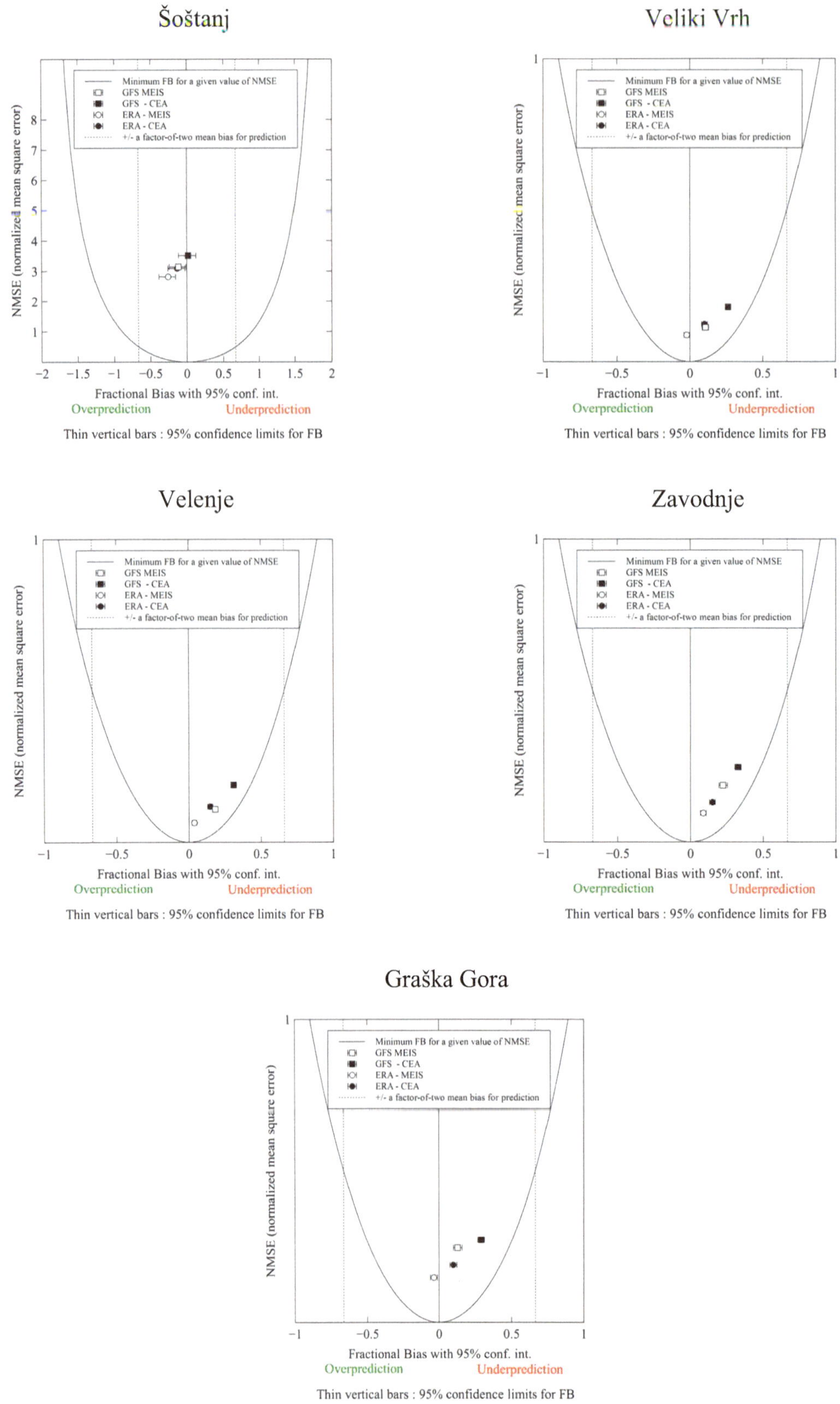

FIG. 32. Statistical analysis of the modelled and measured temperature for each simulation, by station, from 15 March 1991 (07:00) to 5 April 1991 (12:00).

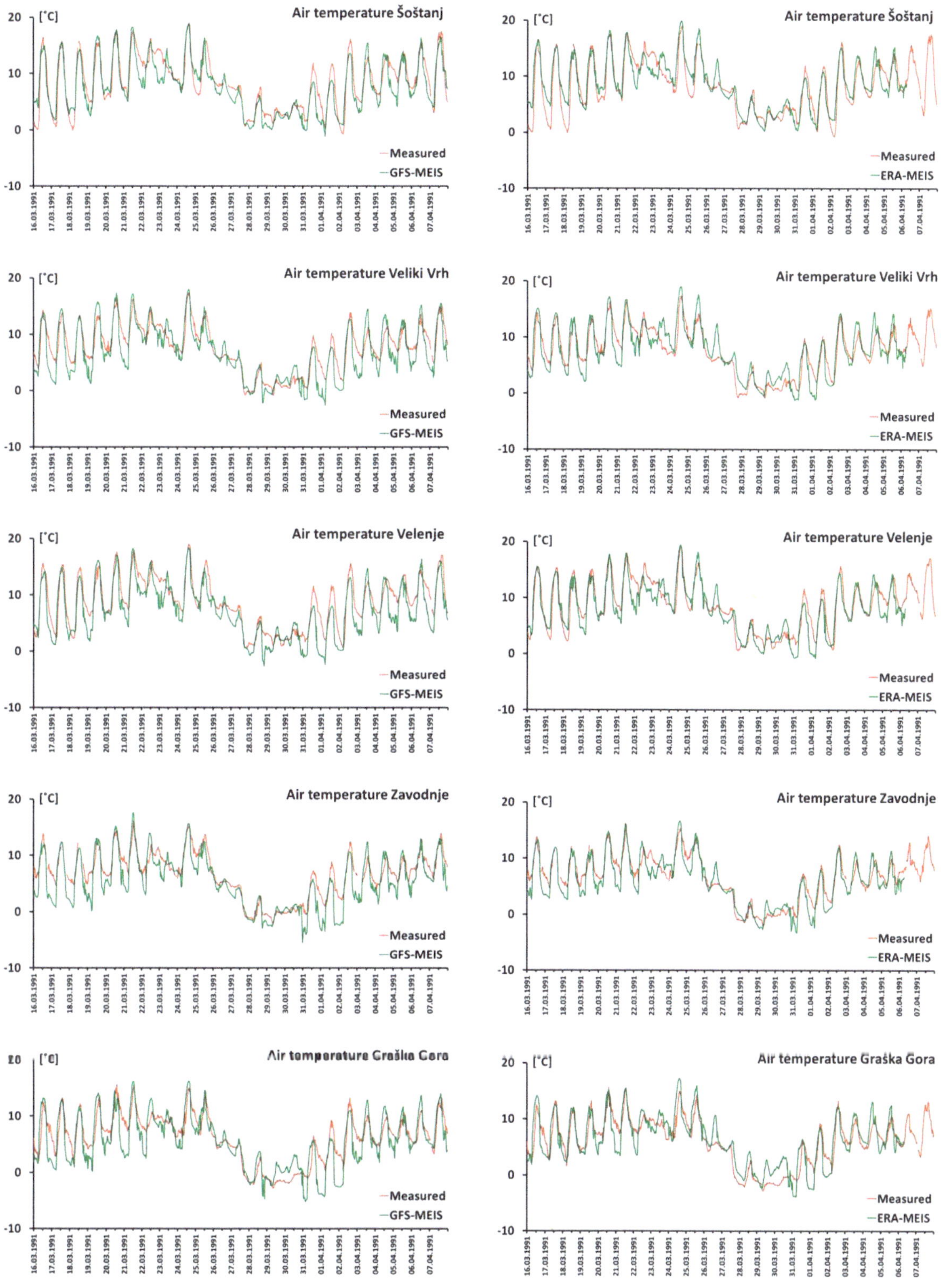

FIG. 33. Timeline of the measured and modelled temperature at five measuring stations (daily from 16 March to 7 April 1991), simulation results from MEIS-1 (left column) and MEIS-2 (right column).

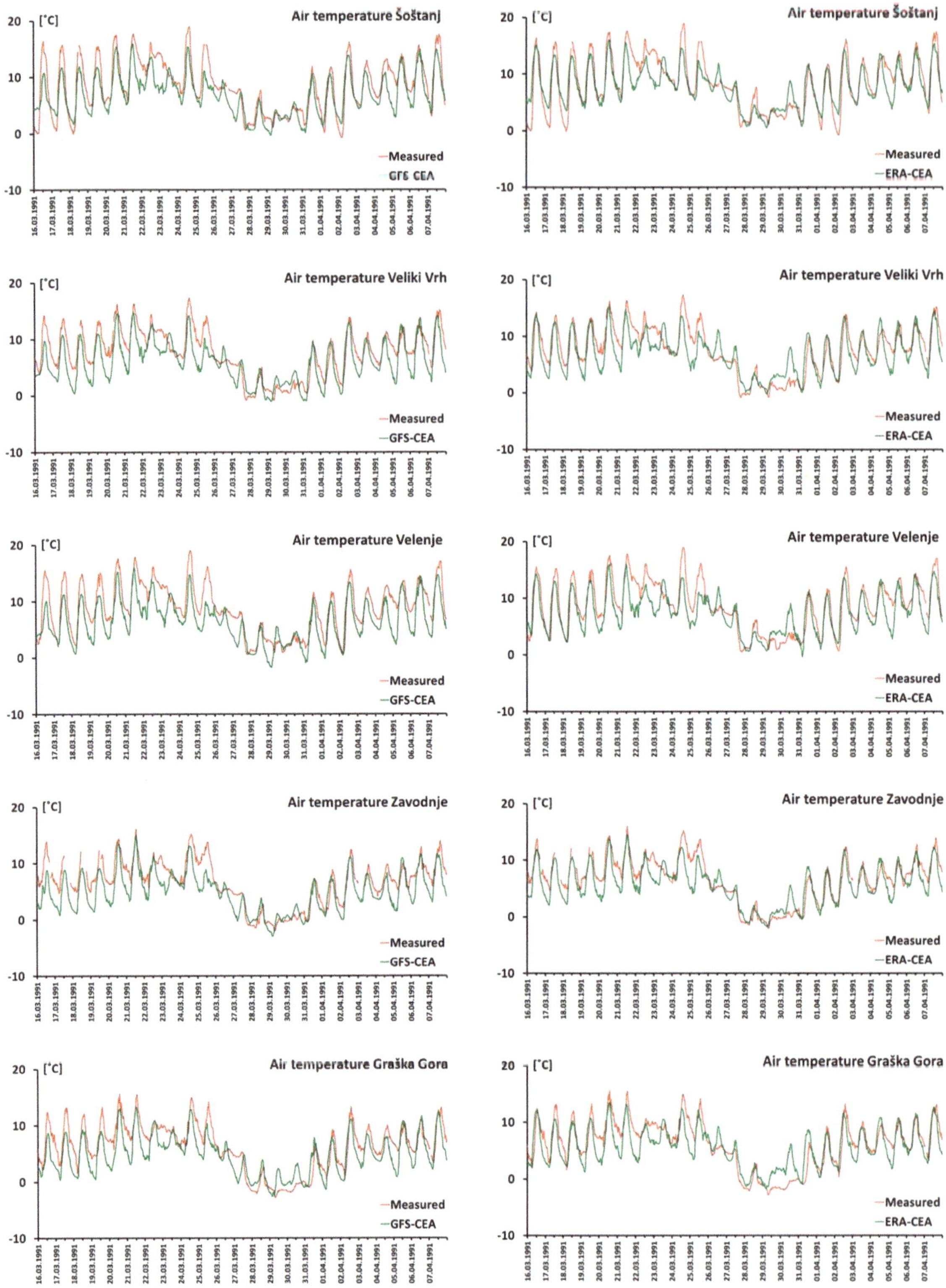

FIG. 34. Timeline of the measured and modelled temperature at five measuring stations (daily from 16 March to 7 April 1991), simulation results from CEA-1 (left column) and CEA-2 (right column).

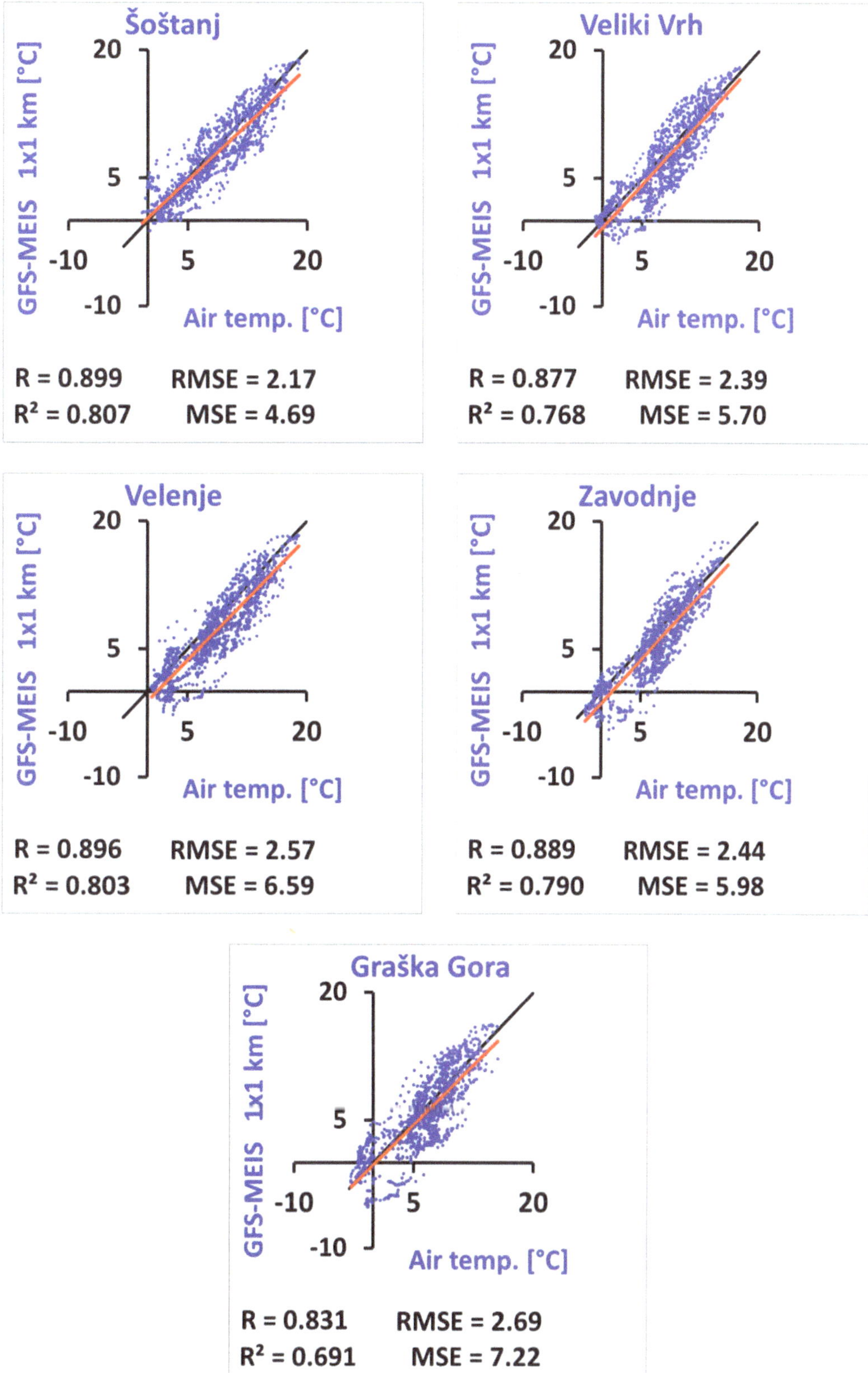

FIG. 35. Scatter plots for measured and modelled temperature at five locations, simulation results from MEIS-1.

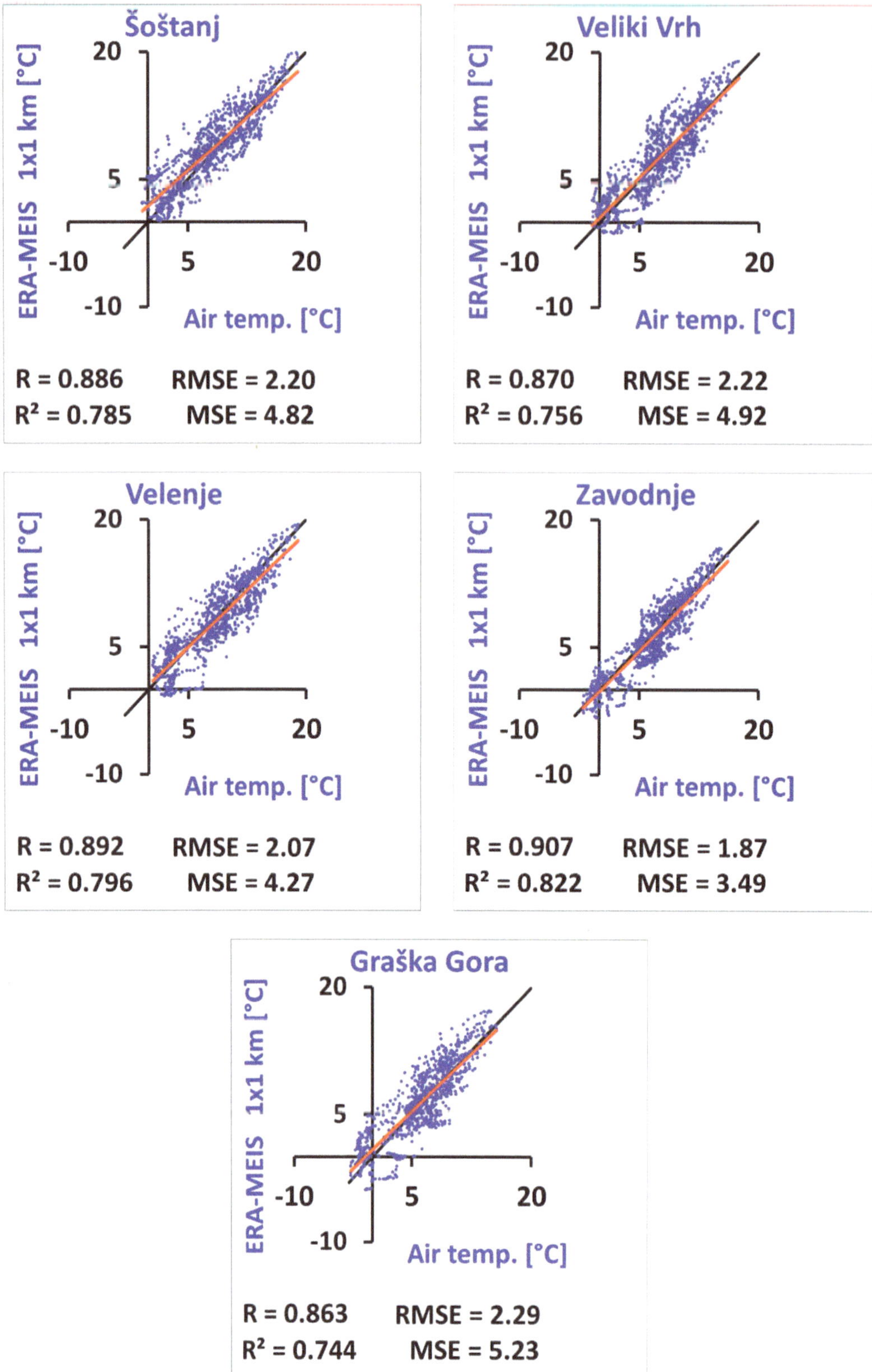

FIG. 36. Scatter plots for measured and modelled temperature at five locations, simulation results from MEIS-2 reproduced from Ref. [15], with permission courtesy of Harmo Initiative).

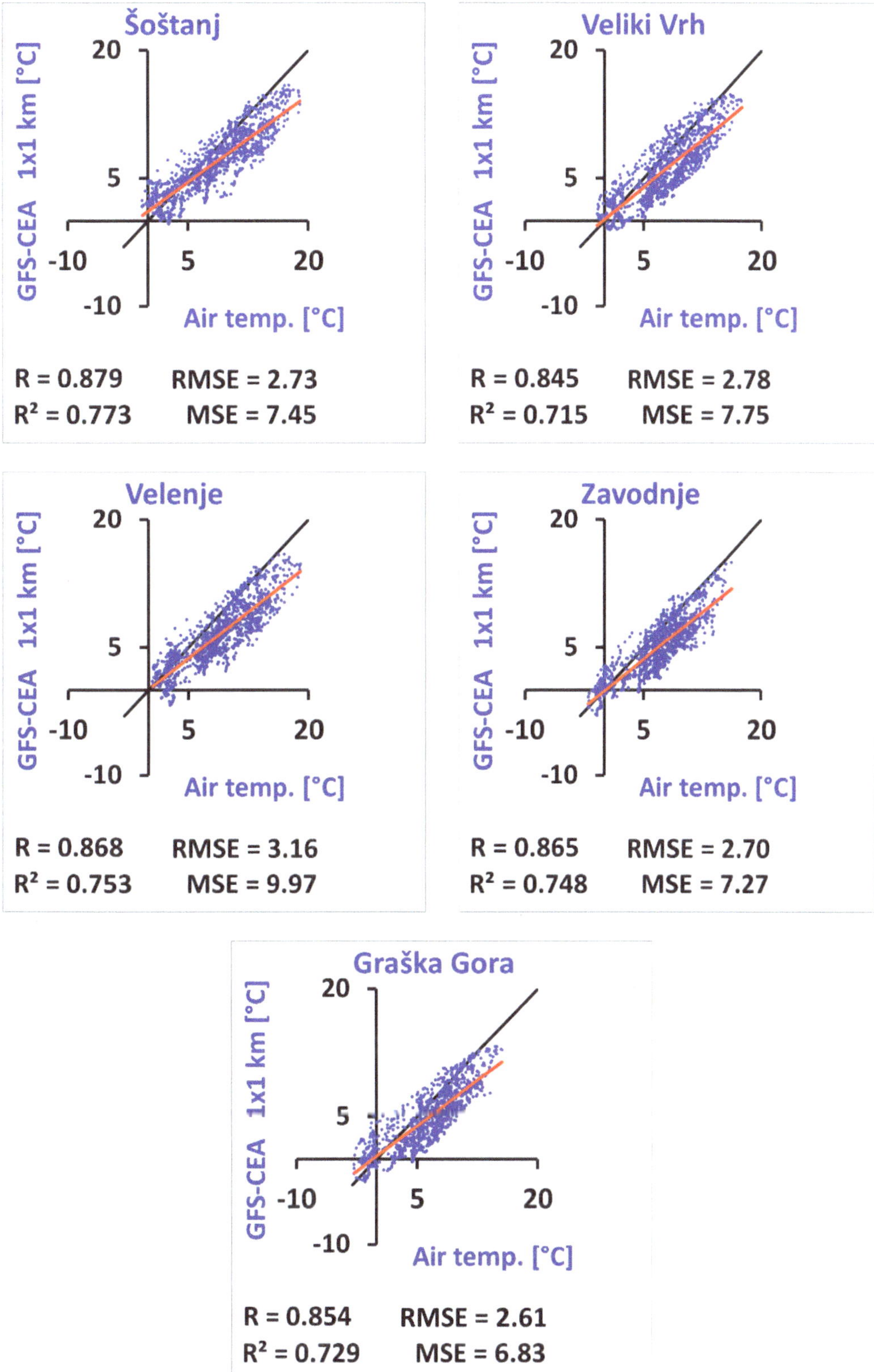

FIG. 37. Scatter plots for measured and modelled temperature at five locations, simulation results from CEA-1 (reproduced from Ref. [15], with permission courtesy of Harmo Initiative).

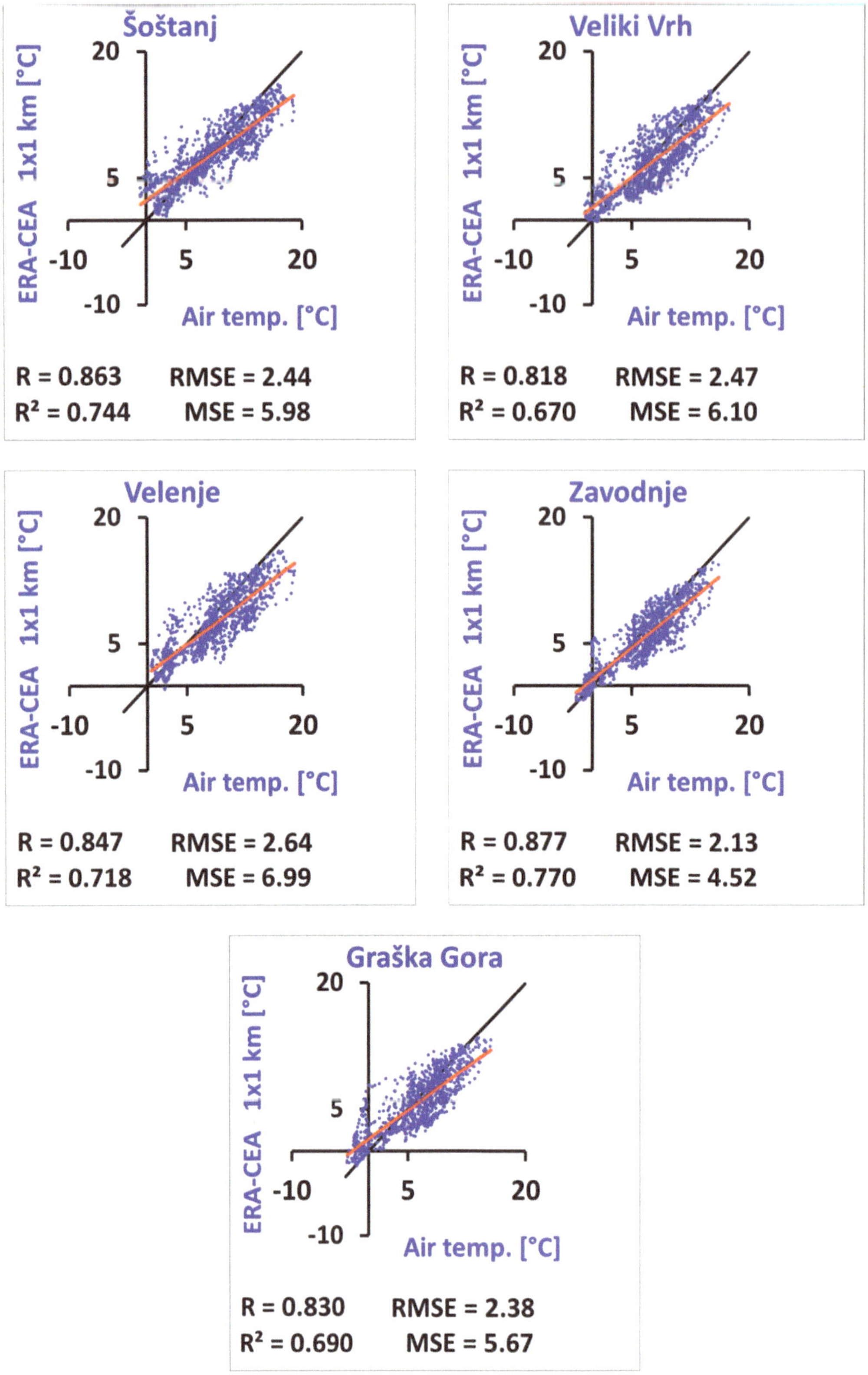

FIG. 38. Scatter plots for measured and modelled temperature at five locations, simulation results from CEA-2.

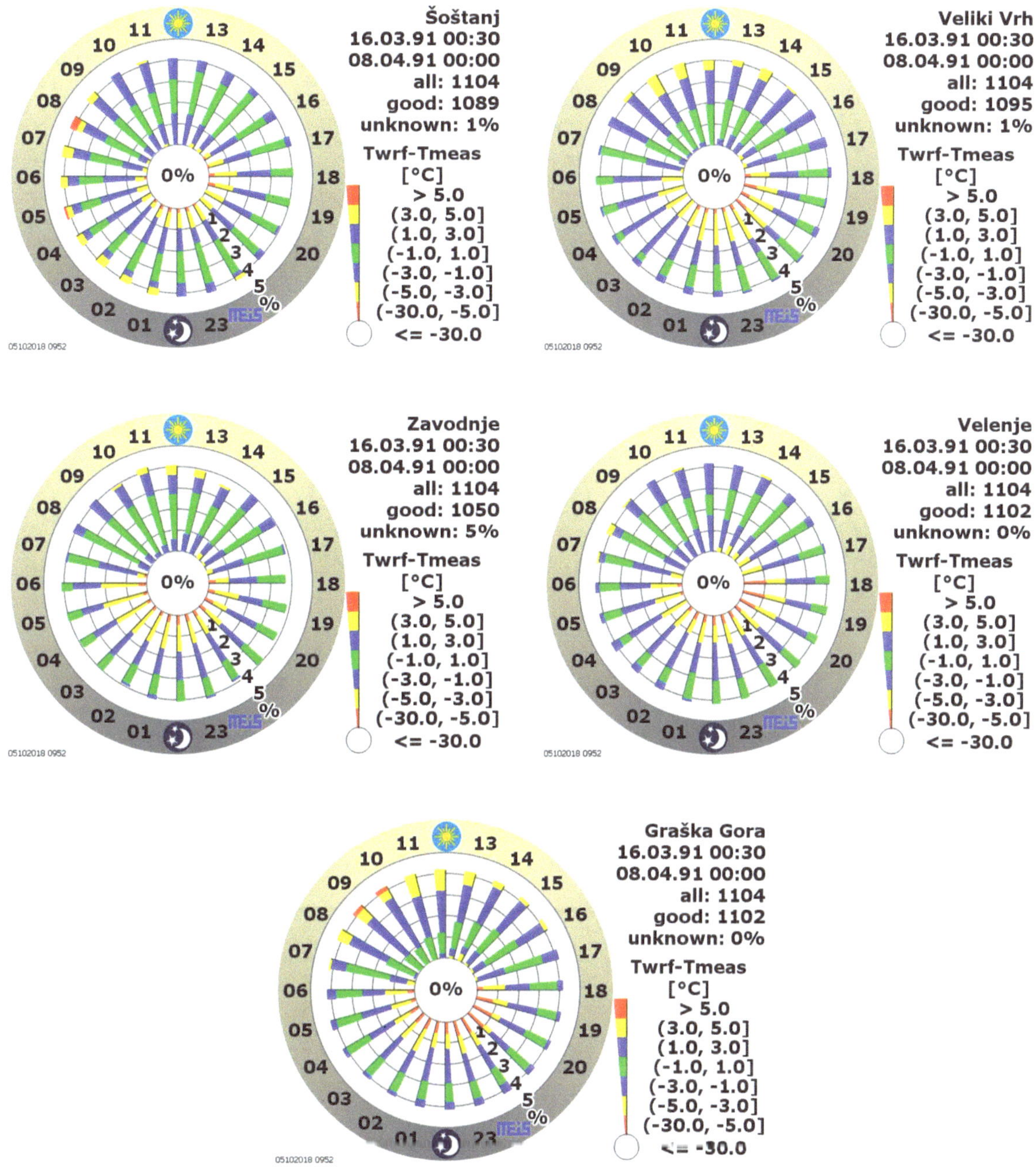

FIG. 39. Daily cycle of errors in temperature forecasts, visualized in a sunflower plot for each of the five locations; simulation results from MEIS-1.

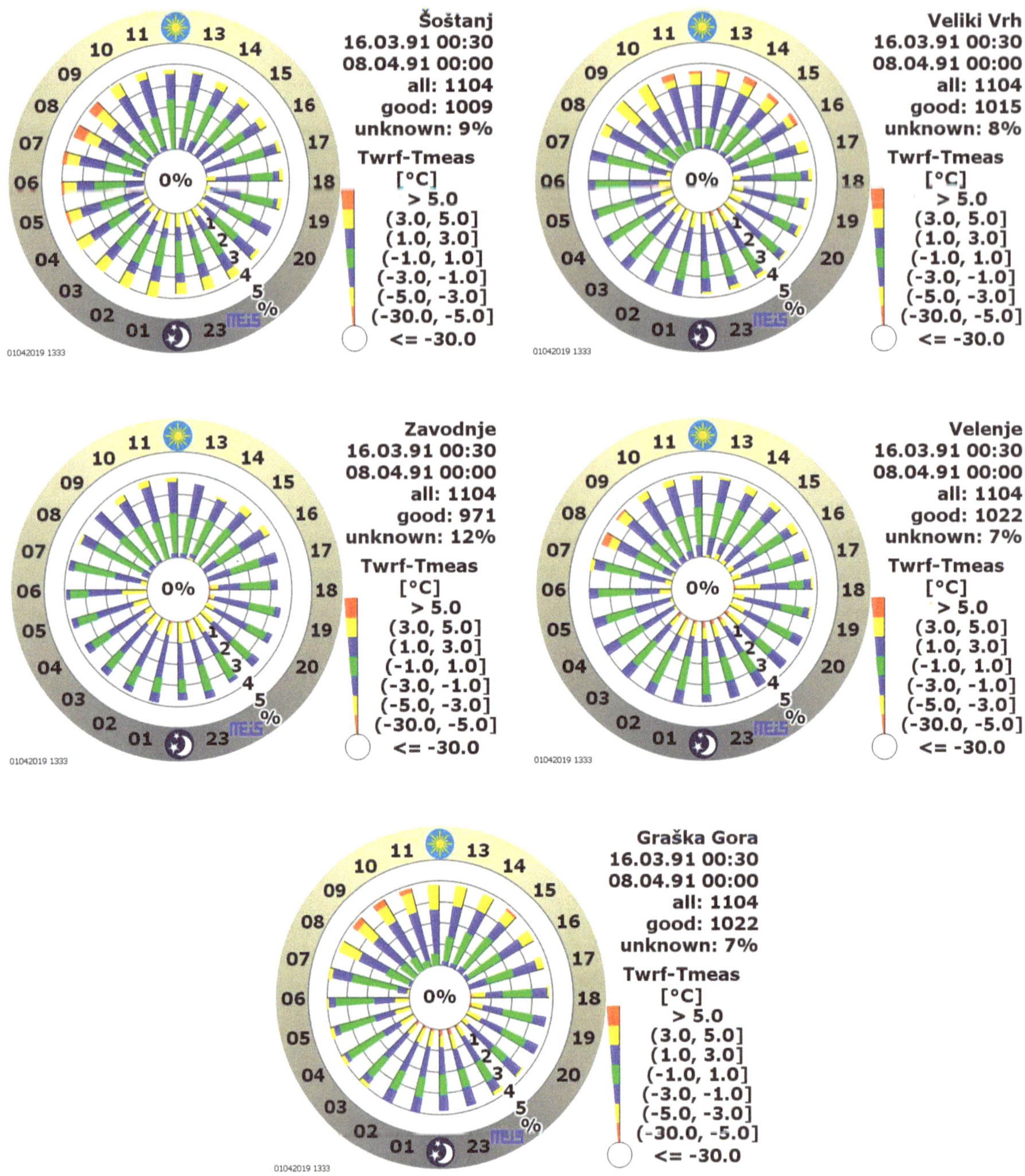

FIG. 40. Daily cycle of errors in temperature forecasts, visualized in a sunflower plot for each of the five locations; simulation results from MEIS-2 reproduced from Ref. [15], with permission courtesy of Harmo Initiative).

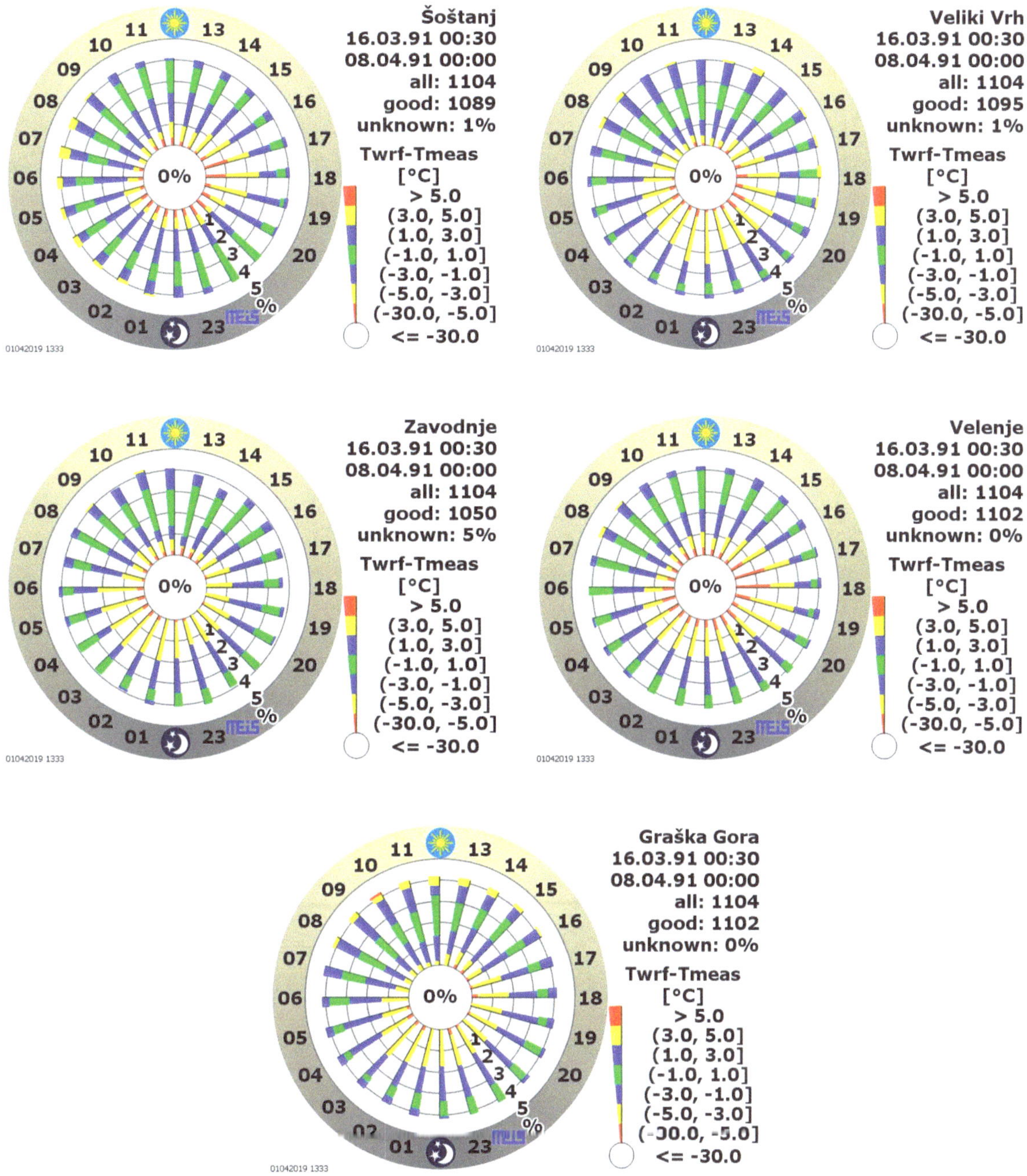

FIG. 41. Daily cycle of errors in temperature forecasts, visualized in a sunflower plot for each of the five locations; simulation results from CEA-1 reproduced from Ref. [15], with permission courtesy of Harmo Initiative).

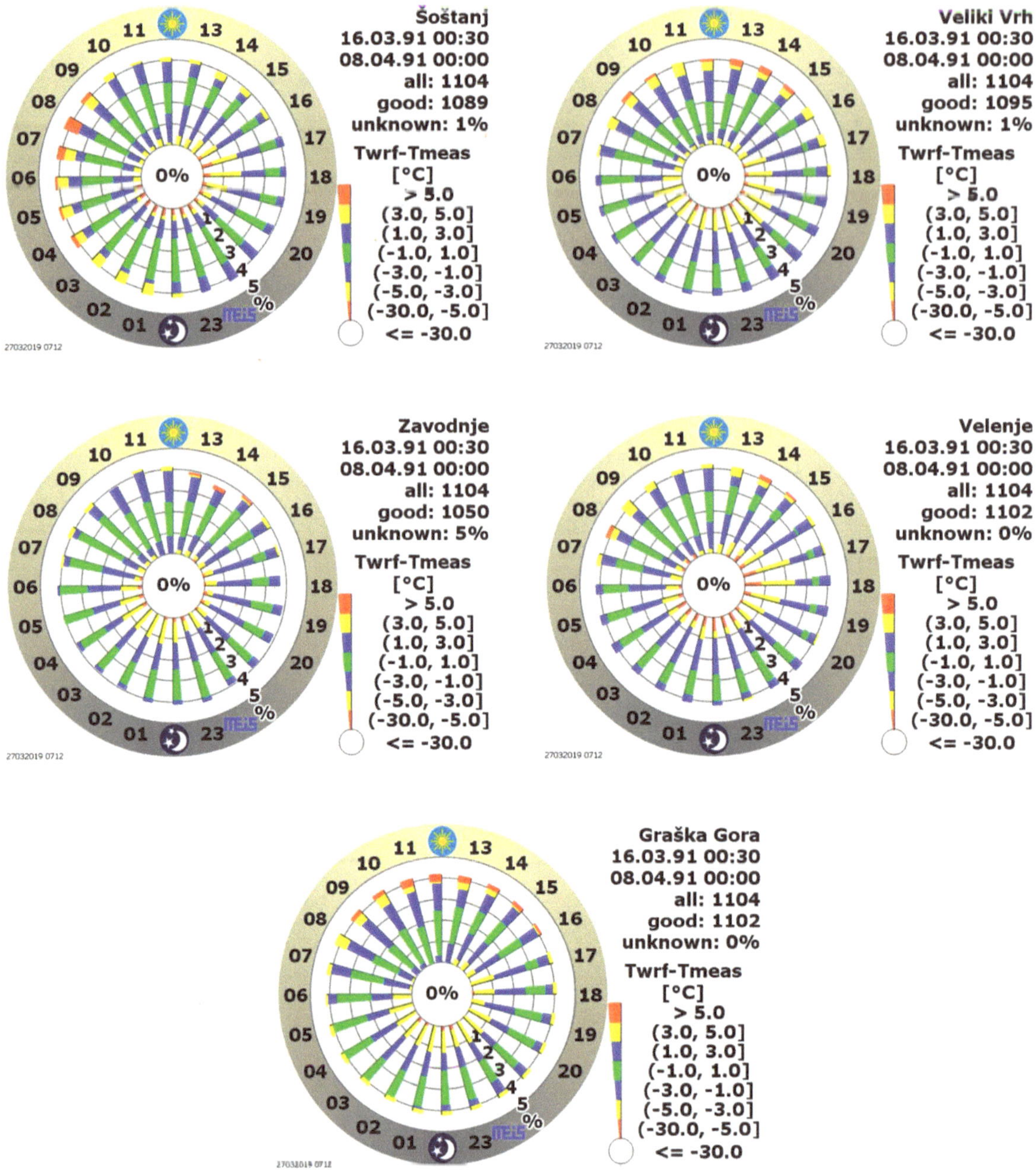

FIG. 42. Daily cycle of errors in temperature forecasts, visualized in a sunflower plot for each of the five locations; simulation results for CEA-2.

5.3.2. Wind

Figure 43 provides a statistical analysis of the predicted and measured wind speeds at each of five meteorological stations for the period 15 March 1991 to 5 April 1991. Comparisons of SODAR measurements and WRF forecasts of the vertical wind profile are provided only for three selected shorter periods (20 March 1991, 30 March 1991 and 2 April 1991; Fig. 44) due to the relatively long validation period. Figures 45–51 show comparisons of the SODAR measurements of vertical wind profile with simulations for 20 March 1991, 30 March 1991, 1 April 1991 and 2 April 1991. The entire time period is visualized with wind rose plots to compare measurements and simulations for the five individual stations (Figs 52 and 53), and for two SODAR levels (Fig. 54).

5.3.3. Global solar radiation

The timeline of global solar radiation and forecasts for the one site where measurements were available have been mapped (Fig. 55). A measurement and forecast sunflower diagram (Fig. 56) and a sunflower diagram with error predictions (Fig. 57) are also provided.

5.3.4. Precipitation

Figure 58 provides the timeline of precipitation measurements and forecasts from each simulation for the one site for which measurements were available.

5.4. DISCUSSION

5.4.1. Air temperature

The weather during the time period of the measurements varied from warm spring temperatures to a period of cool air and snow, followed by a warming trend. This variability provides a good opportunity to test forecasts of temperature. Figures 32–38 show that forecasts matched measurements reasonably well. time (Figs 39–42), with the models predicting lower temperatures than were actually measured. Daytime forecasts were significantly more accurate (except for the Šoštanj station between 07:00 and 08:00). In general, the largest differences (nighttime or daytime) were seen for the Graška Gora station. Significant differences are seen for night.

5.4.2. Wind

Comparison of predicted and measured wind profiles (Figs 44–51) shows that predicted wind speeds are generally too high or, at best, roughly accurate (e.g. 2 April 1991). Predicted wind directions were sometimes for the layers nearest to the ground. Predicted wind roses (Figs 52–54) matched measurements well for the Veliki Vrh and Graška Gora stations (located atop their respective hills) and both SODAR levels, but predictions were much poorer for the Šoštanj and Velenje stations (at the bottom of the valley) and in Zavodnje (located on the slope).

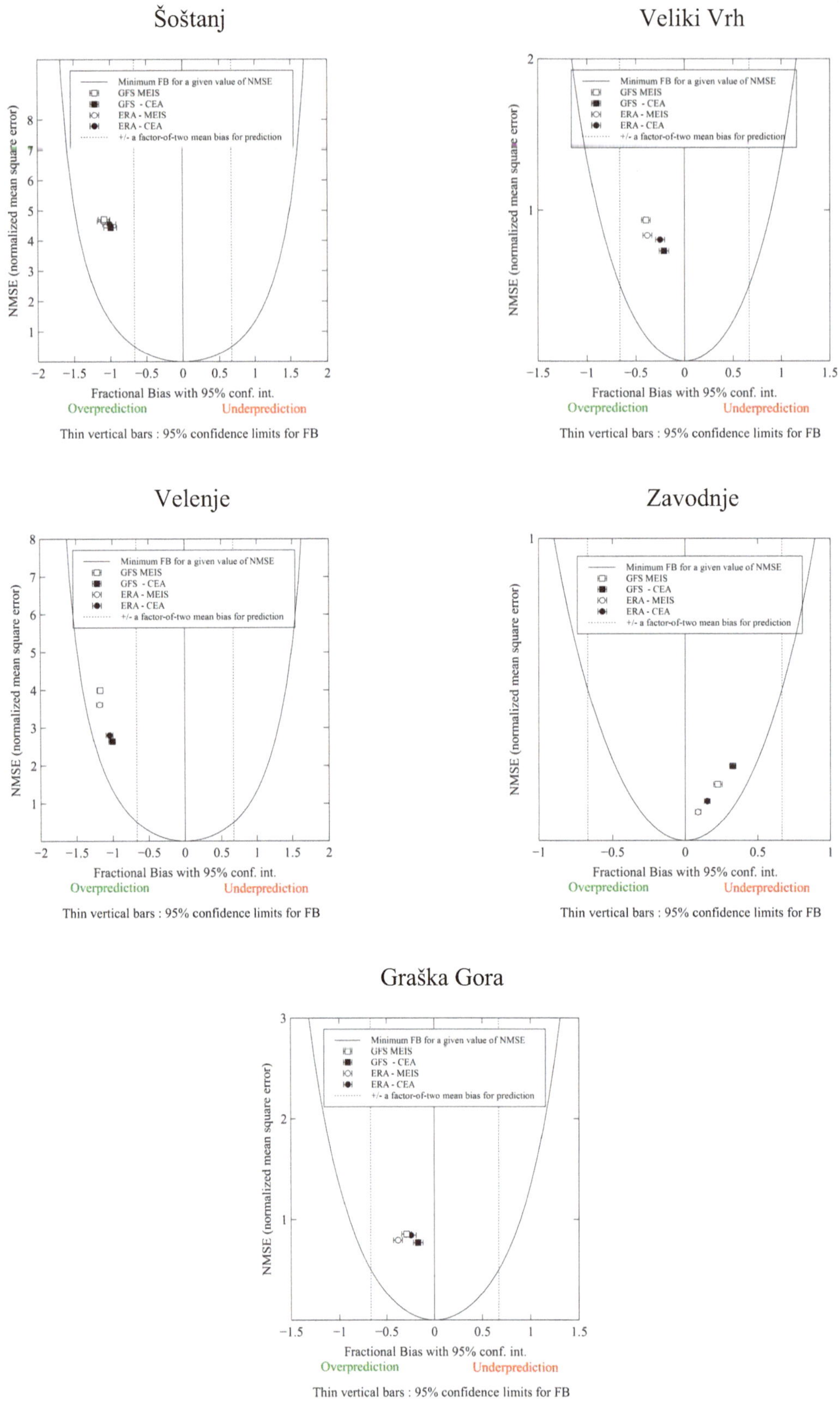

FIG. 43. Statistical analysis of the modelled and measured wind speed for each simulation, by station, from 15 March 1991 (07:00) to 5 April 1991 (12:00).

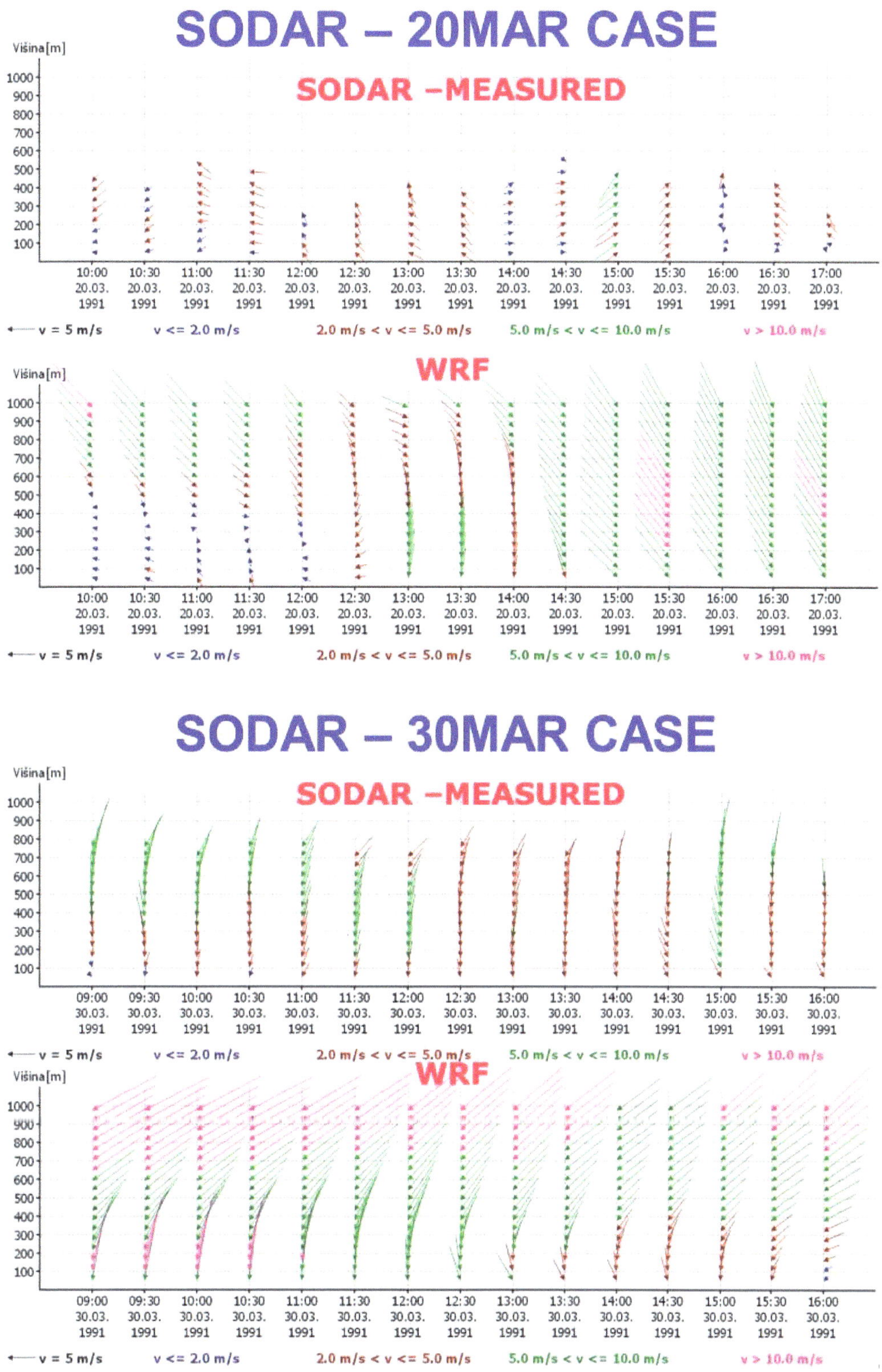

FIG. 44. Comparison of vertical wind profiles for three selected periods (20 March 1991, 30 March 1991, 2 April 1991). Visualization of SODAR measurements and WRF forecasts, as a function of height above the ground (m) on y-axis. MEIS-1 simulation result.

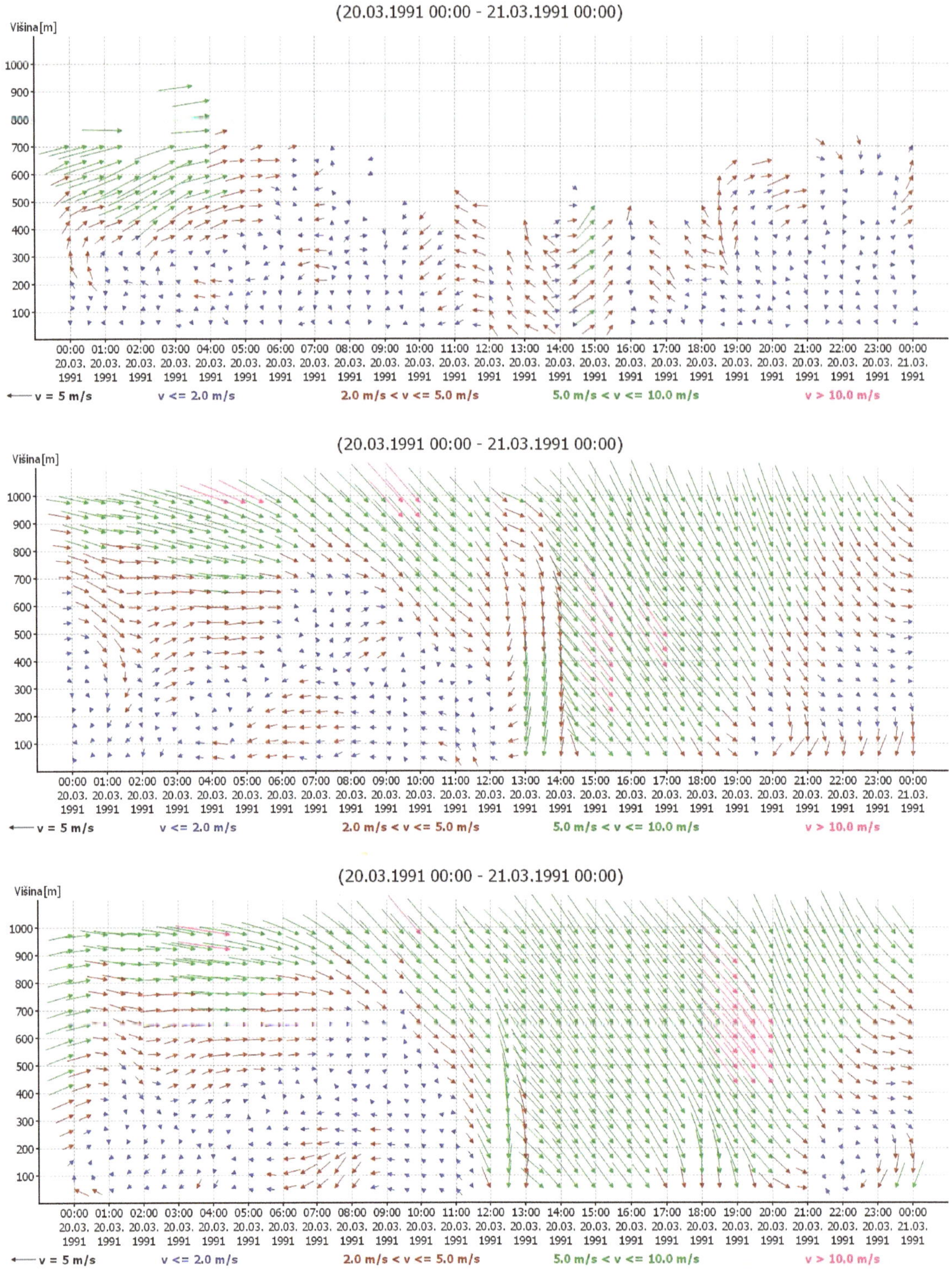

FIG. 45. Comparison of vertical wind profiles for 20 March 1991 as a function of height above the ground (m)- y-axis. Visualization of SODAR measurements (top) and simulation results from MEIS-1(middle) and MEIS-2 (bottom).

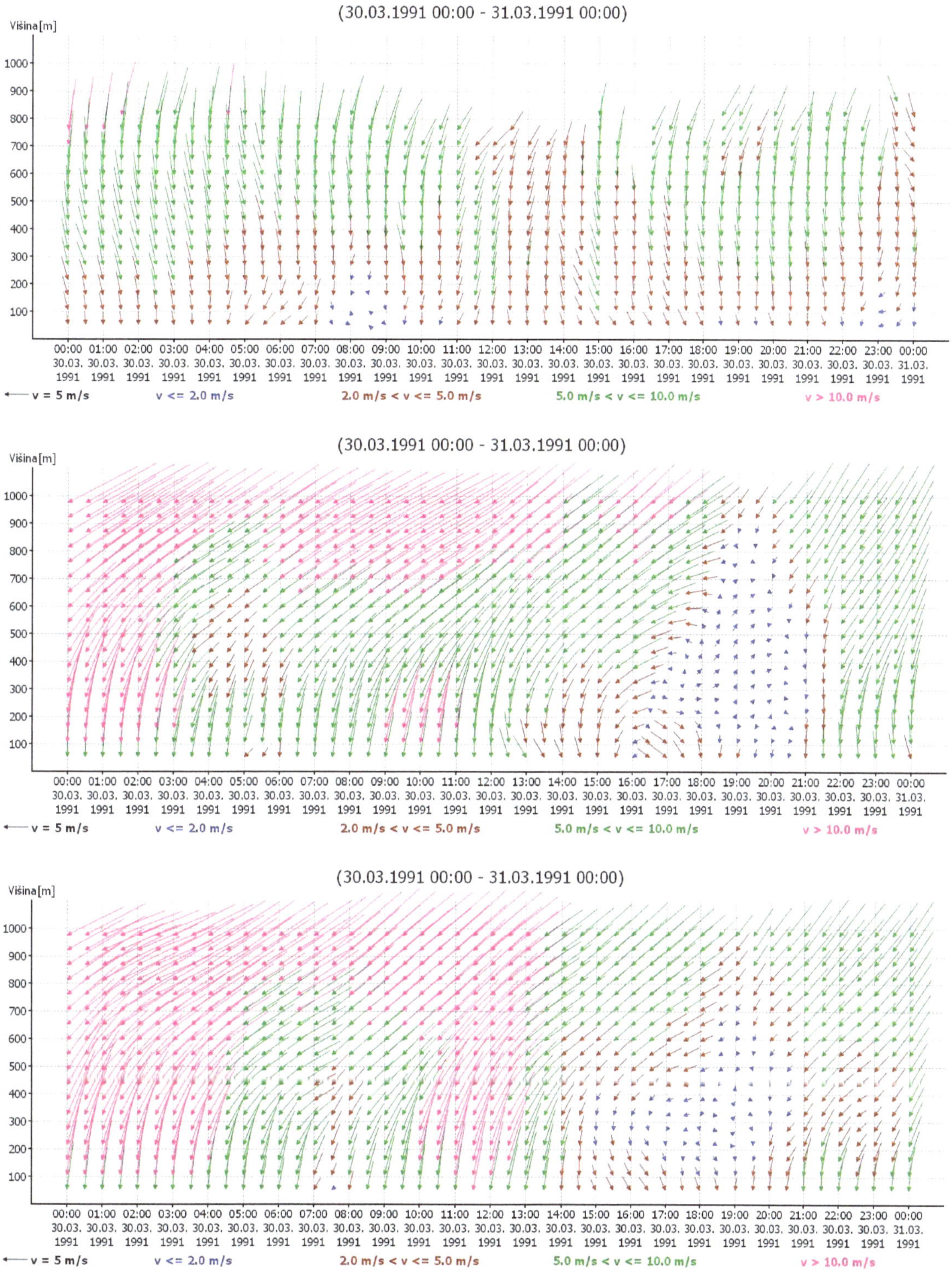

FIG. 46. Comparison of vertical wind profiles for 30 March 1991 as a function of height above the ground (m)- y-axis. Visualization of SODAR measurements (top) and simulation results from MEIS-1(middle) and MEIS-2 (bottom).

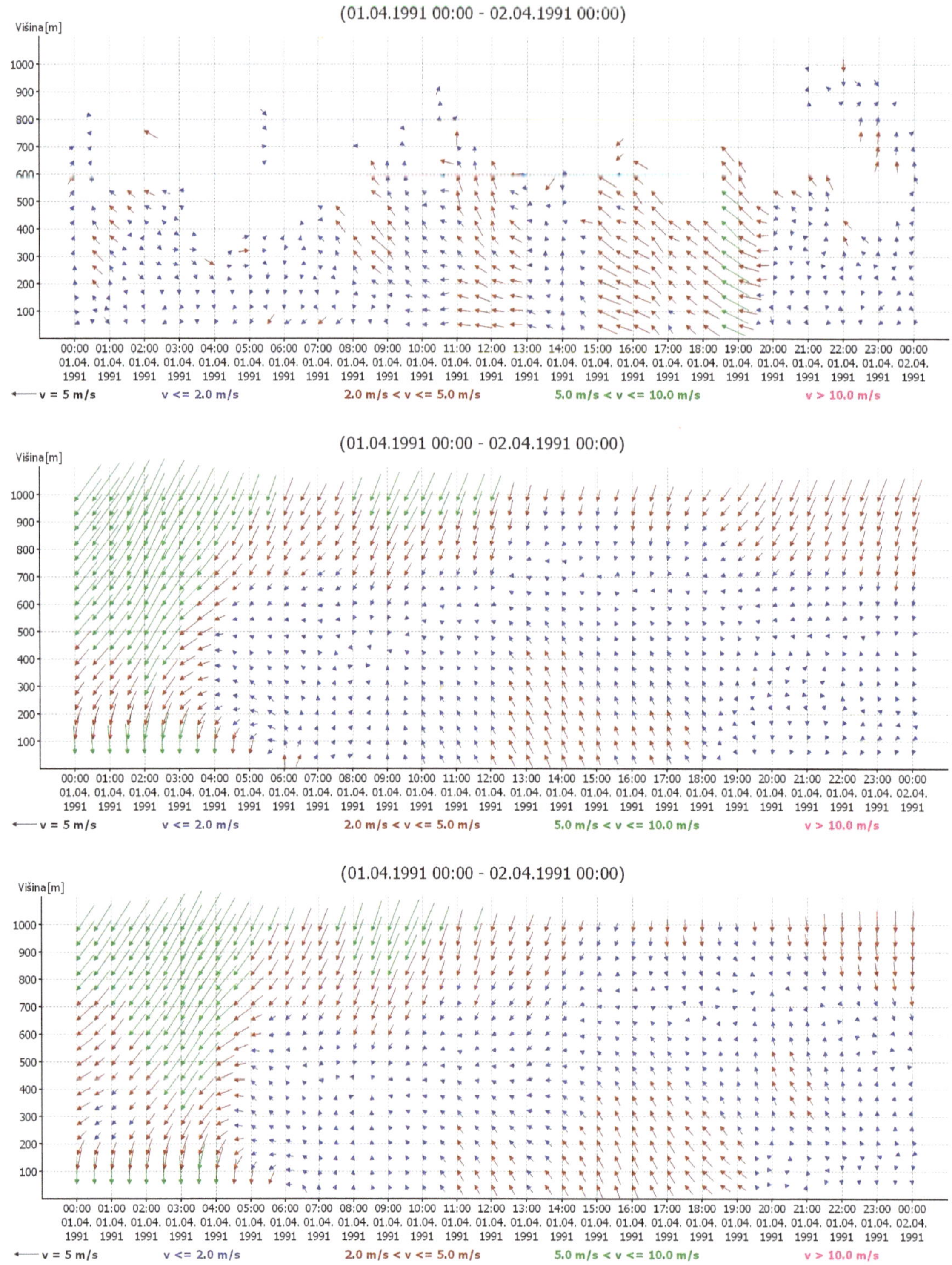

FIG. 47. Comparison of vertical wind profiles for 1 April 1991 as a function of height above the ground (m)- y-axis. Visualization of SODAR measurements (top) and simulation results from MEIS-1(middle) and MEIS-2 (bottom).

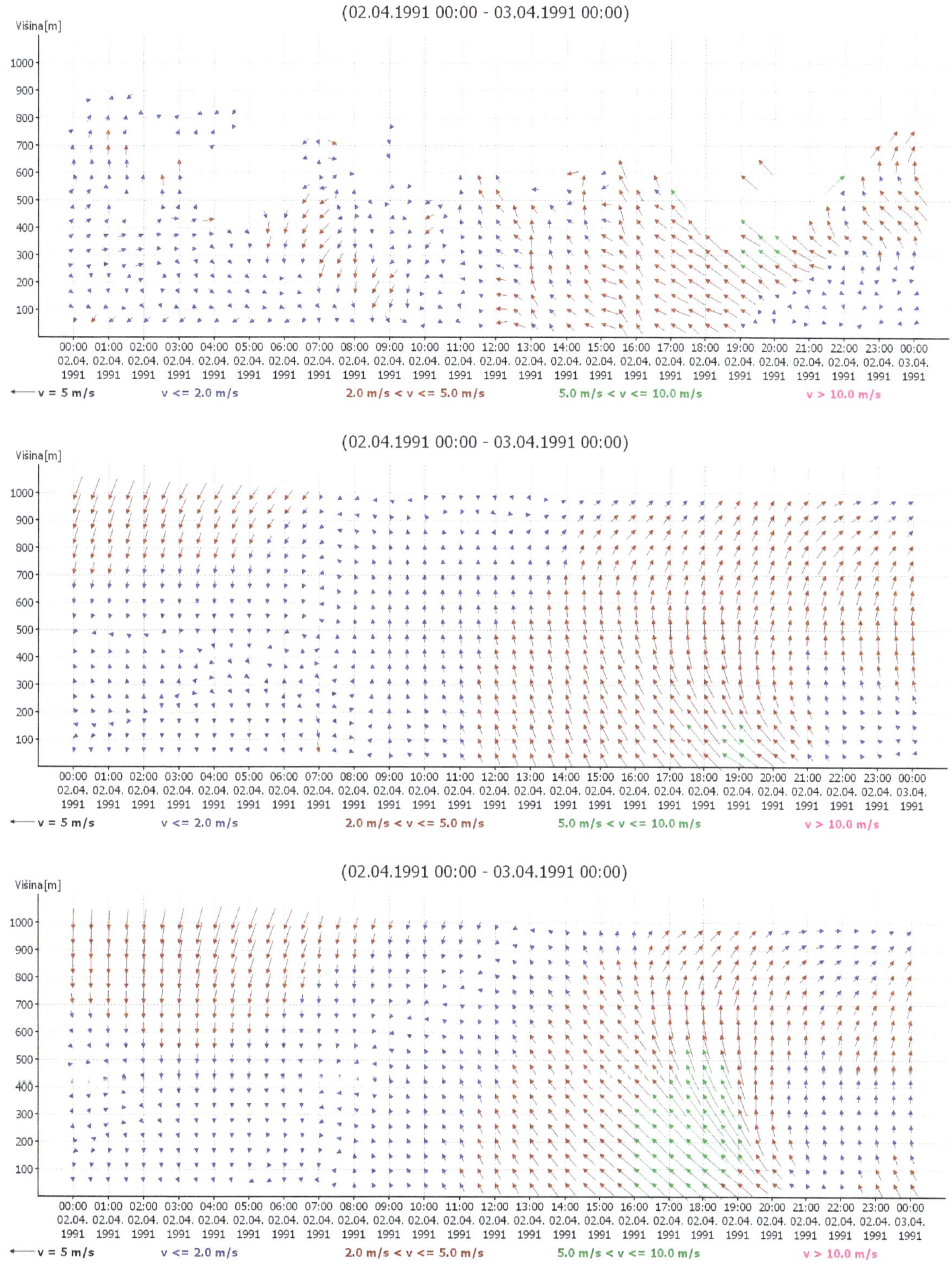

FIG. 48. Comparison of vertical wind profiles for 2 April 1991 as a function of height above the ground (m)- y-axis. Visualization of SODAR measurements (top) and simulation results from MEIS-1(middle) and MEIS-2 (bottom).

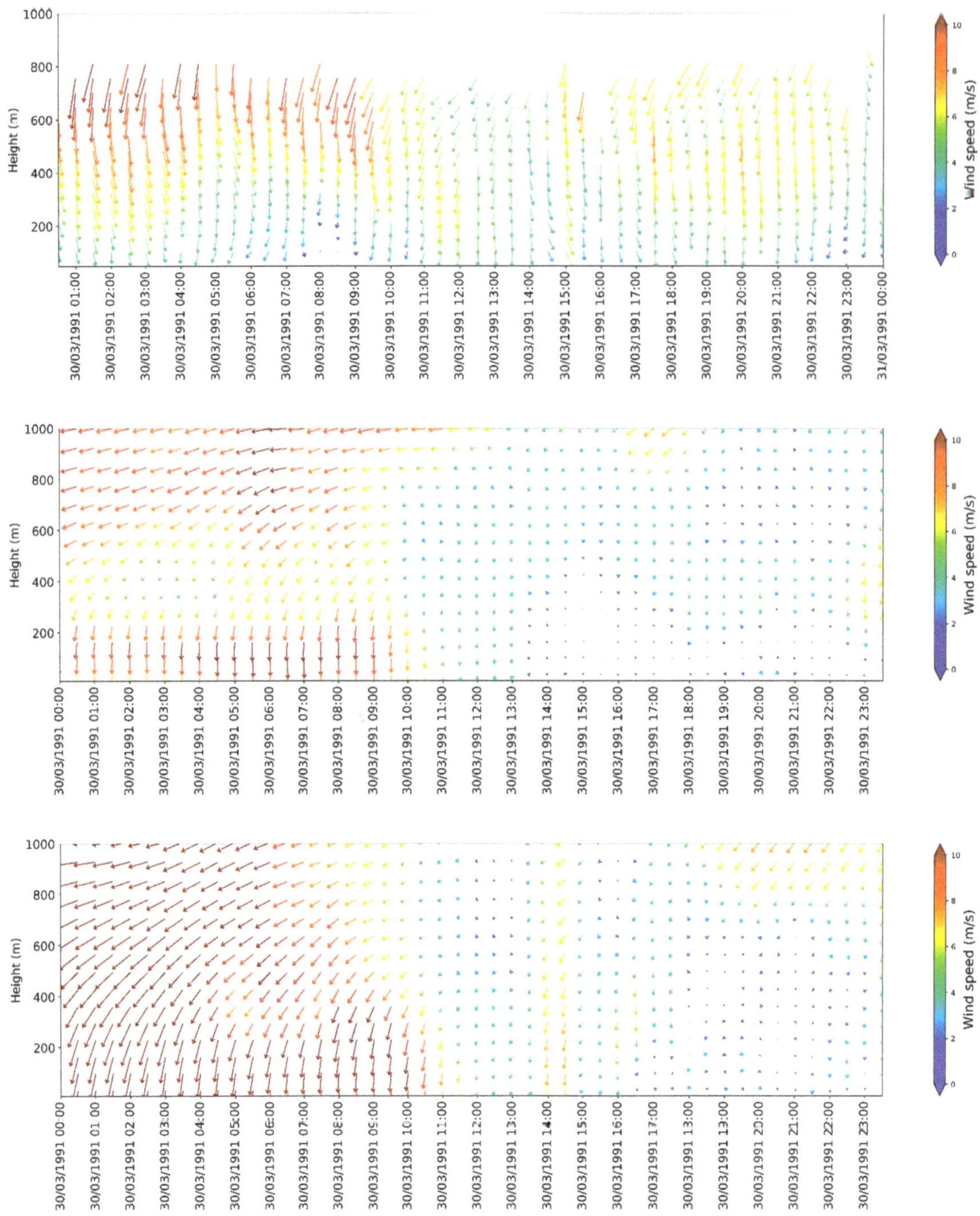

FIG. 49. Comparison of vertical wind profiles for 30 March 1991. Visualization of SODAR measurements (top) and simulation results from CEA-1(middle) and CEA-2 (bottom).

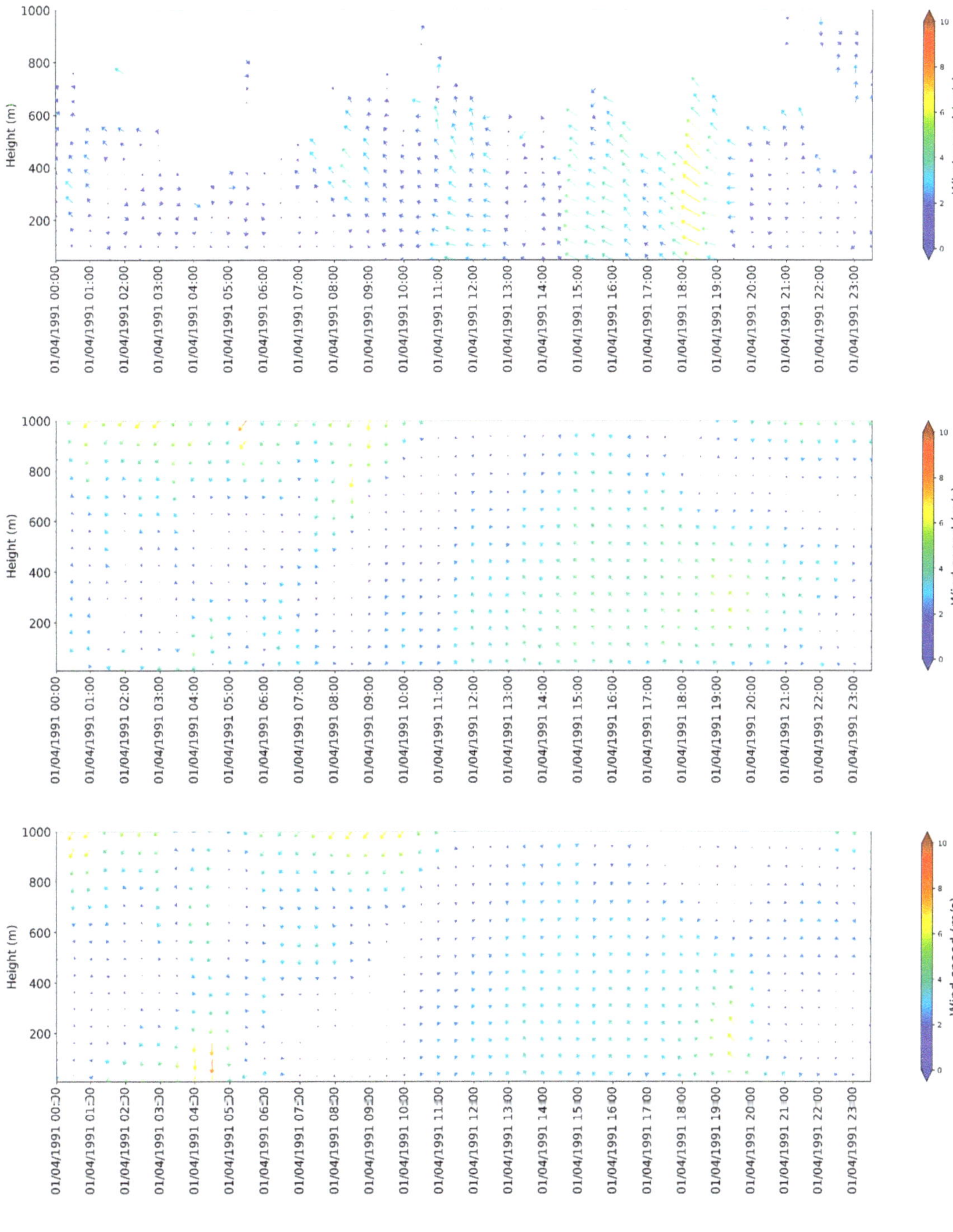

FIG. 50. Comparison of vertical wind profiles for 1 April 1991. Visualization of SODAR measurements (top) and simulation results from CEA-1(middle) and CEA-2 (bottom).

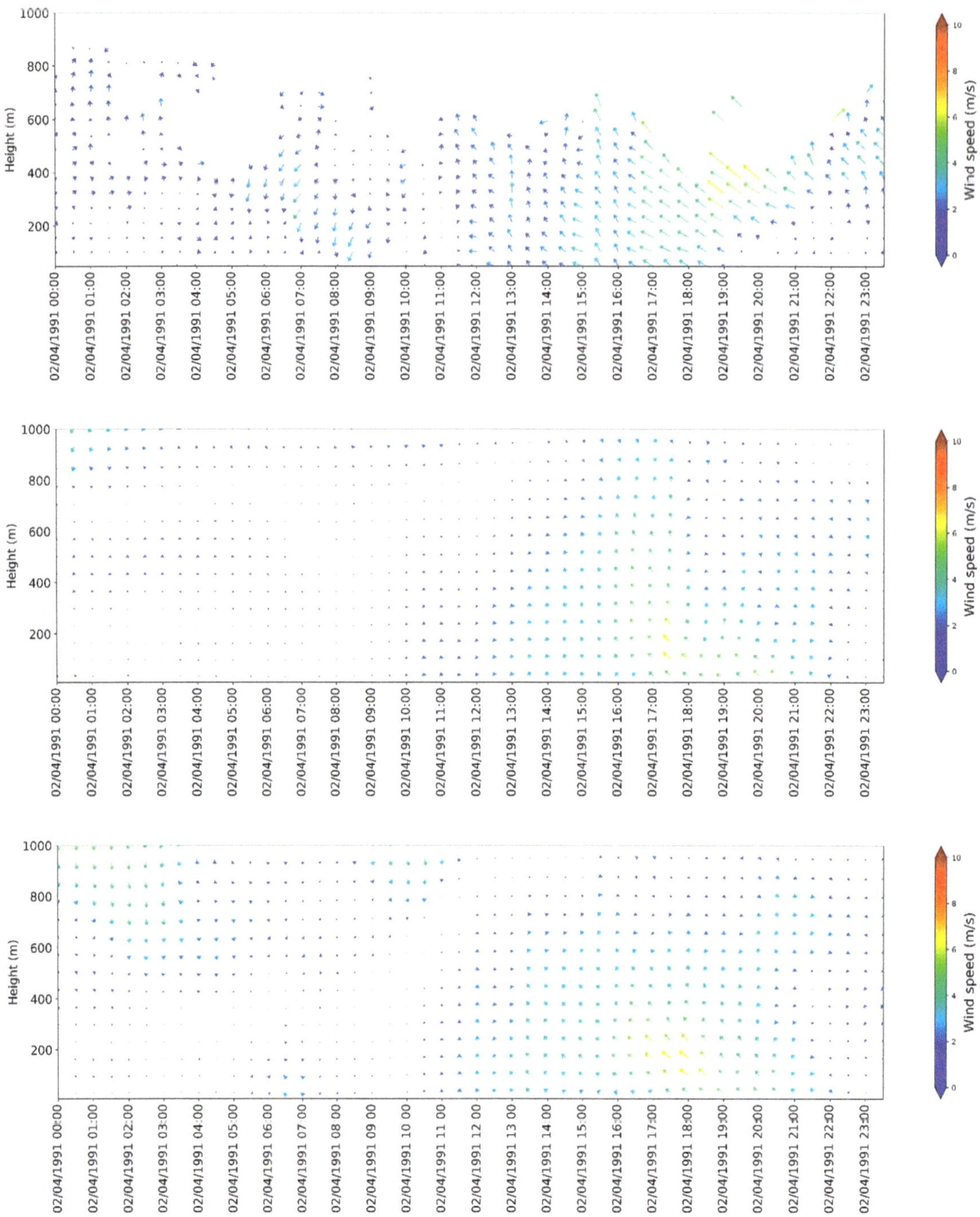

FIG. 51. Comparison of vertical wind profiles for 2 April 1991. Visualization of SODAR measurements (top) and simulation results from CEA-1(middle) and CEA-2 (bottom).

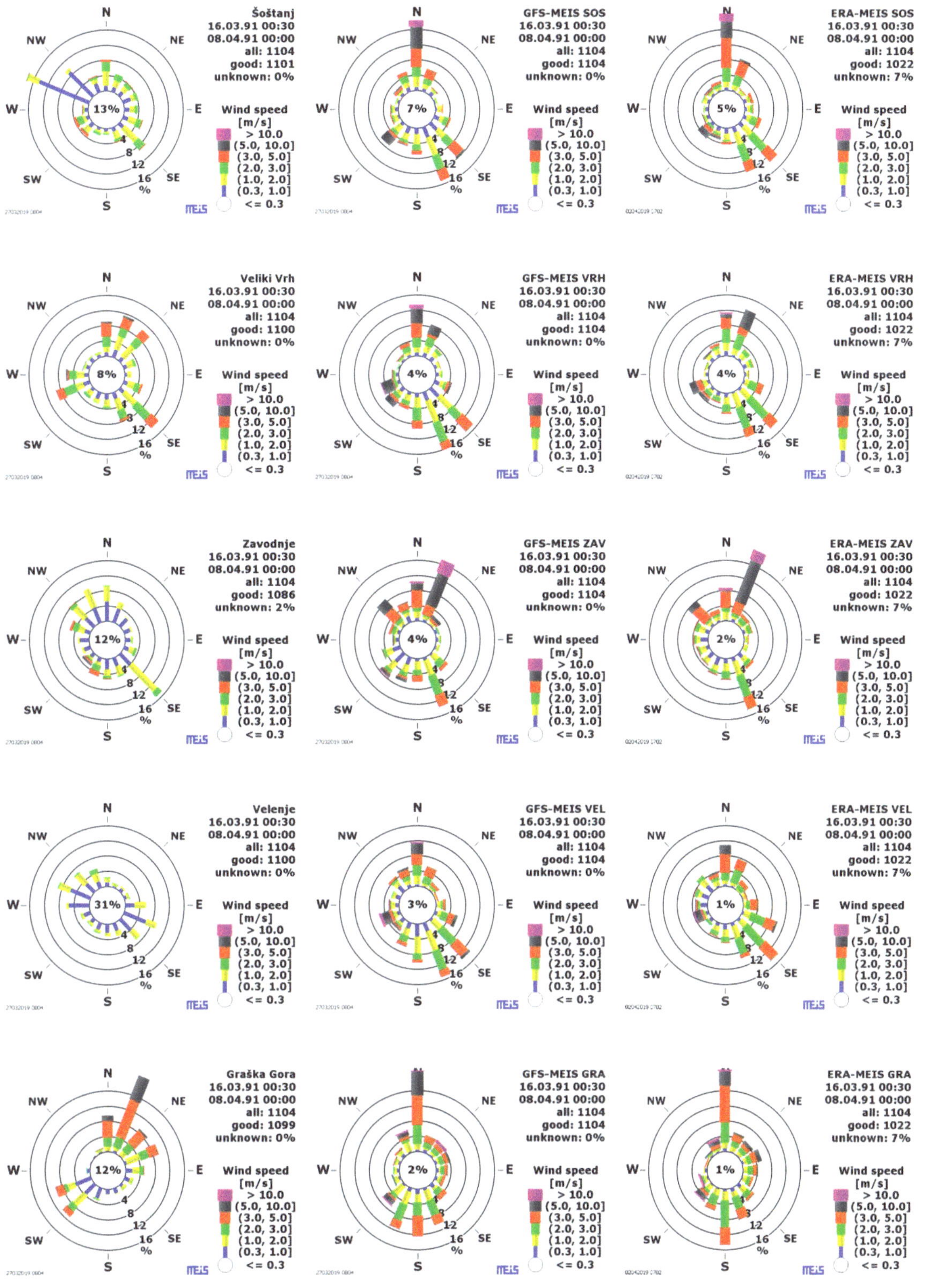

FIG. 52. Wind comparison at five locations (Šoštanj, Veliki Vrh, Zavodnje, Velenje and Graška Gora). Wind roses of measurements for ground level stations (left column) and WRF forecasts from the MEIS-1 and MEIS-2 simulations (middle and right columns) reproduced from Ref. [15], with permission courtesy of Harmo Initiative).

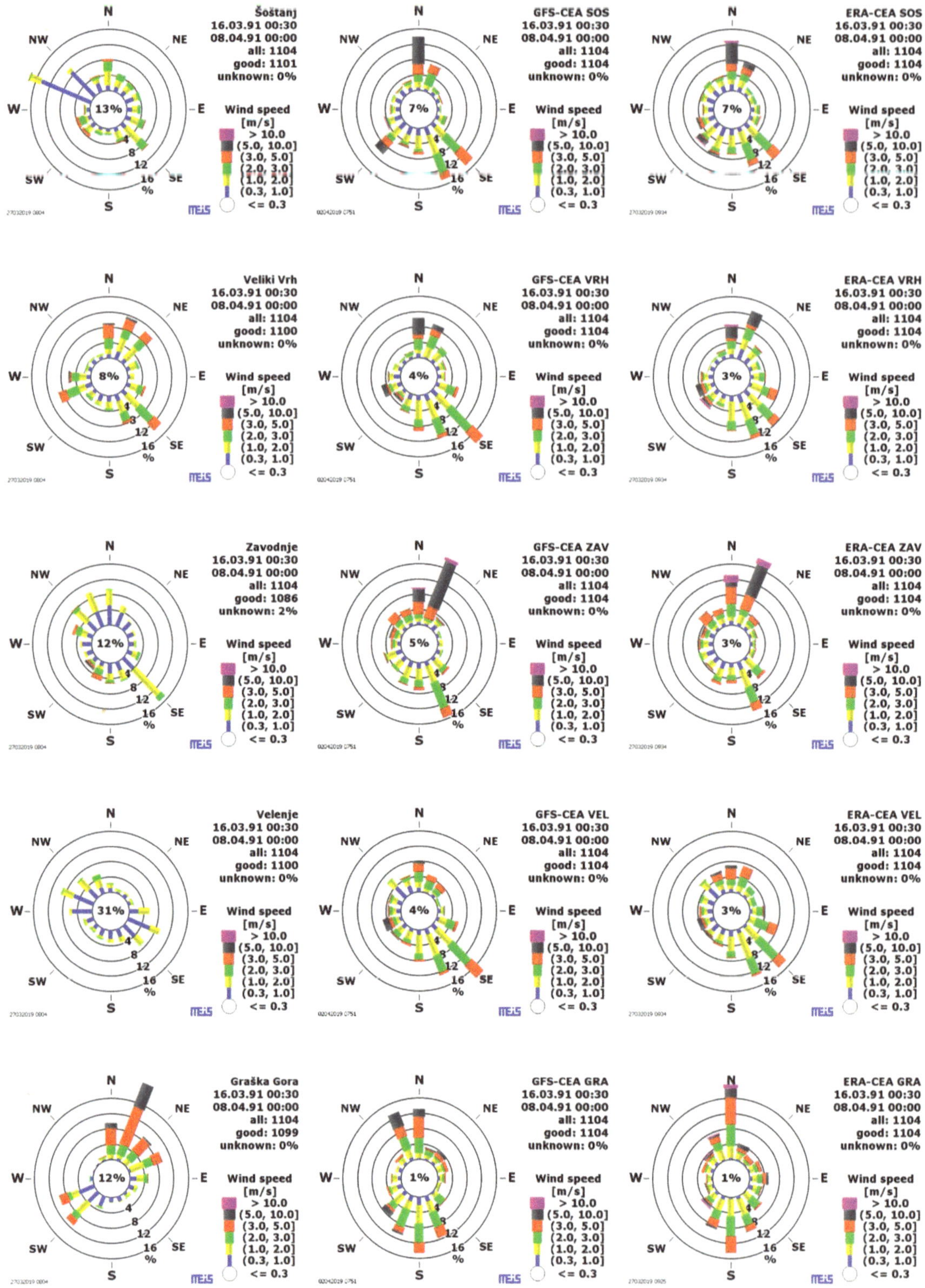

FIG. 53. Wind comparison at five locations (Šoštanj, Veliki Vrh, Zavodnje, Velenje and Graška Gora). Wind roses of measurements for ground level stations (left column) and WRF forecasts from the CEA-1 and CEA-2 simulations (middle and right columns) reproduced from Ref. [15], with permission courtesy of Harmo Initiative).

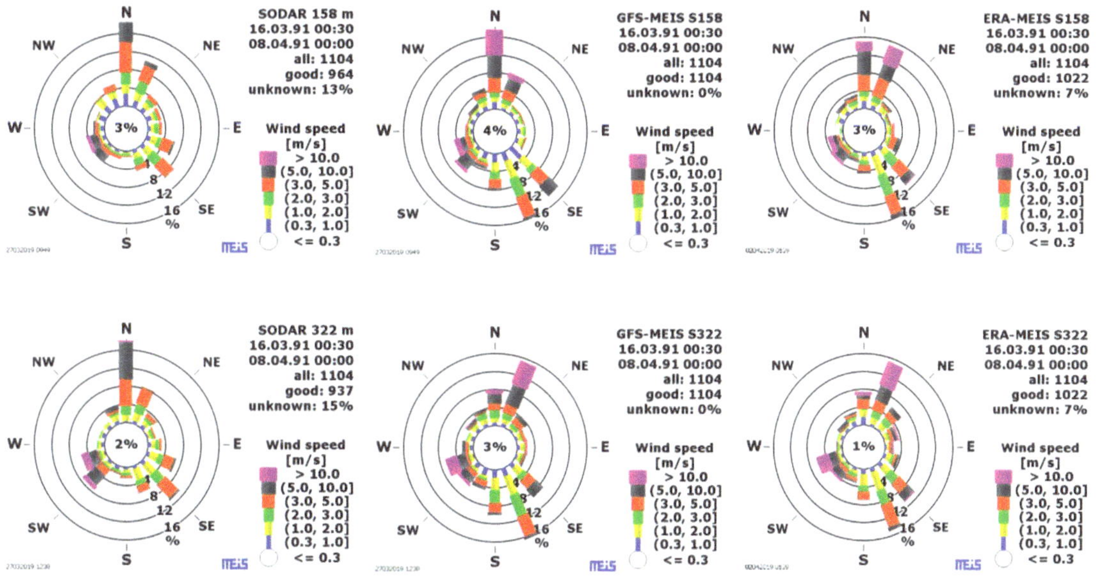

FIG. 54. Wind comparison at SODAR location for two levels at 158 m (top row) and 322 m (bottom row) height above the ground. Wind roses of measurements for ground level stations (left column) and WRF forecasts from the MEIS-1 and MEIS-2 simulations (middle and right columns reproduced from Ref. [15], with permission courtesy of Harmo Initiative).

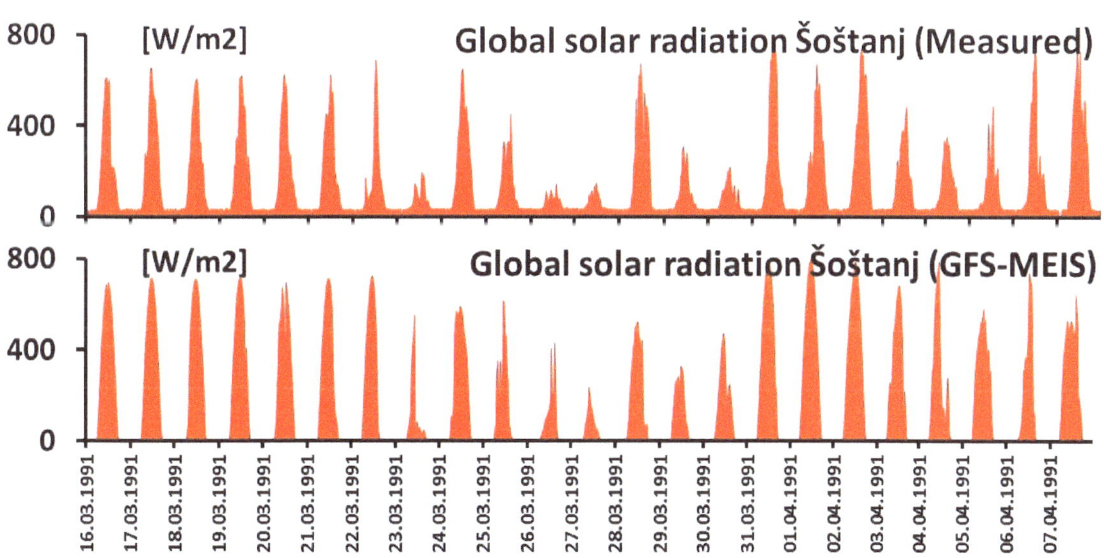

FIG. 55. Timeline of the global solar radiation, measured (top) and simulation results from MEIS-1 (bottom) reproduced from Ref. [15], with permission courtesy of Harmo Initiative).

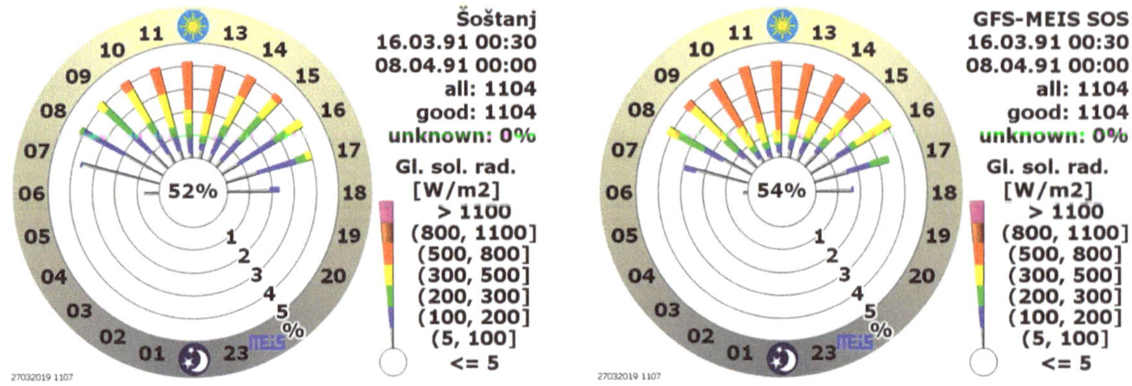

FIG. 56. Analysis of the daily cycle of the global solar radiation, visualized in a sunflower plot, showing measurements (left) and simulation results from MEIS-1 (right) reproduced from Ref. [15], with permission courtesy of Harmo Initiative).

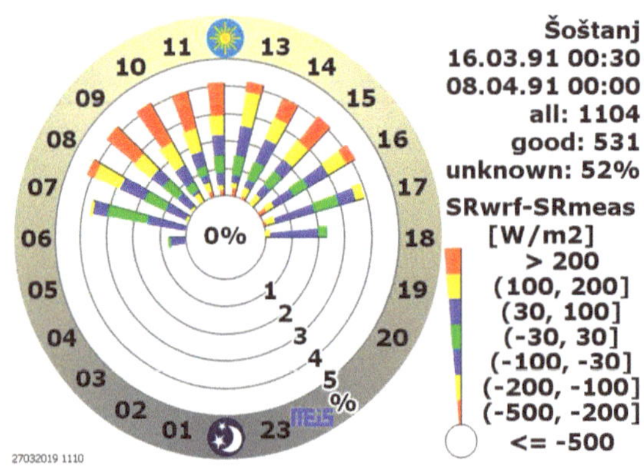

FIG. 57. Analysis of the daily cycle of error predictions for global solar radiation, visualized in a sunflower plot; simulation results from MEIS-1.

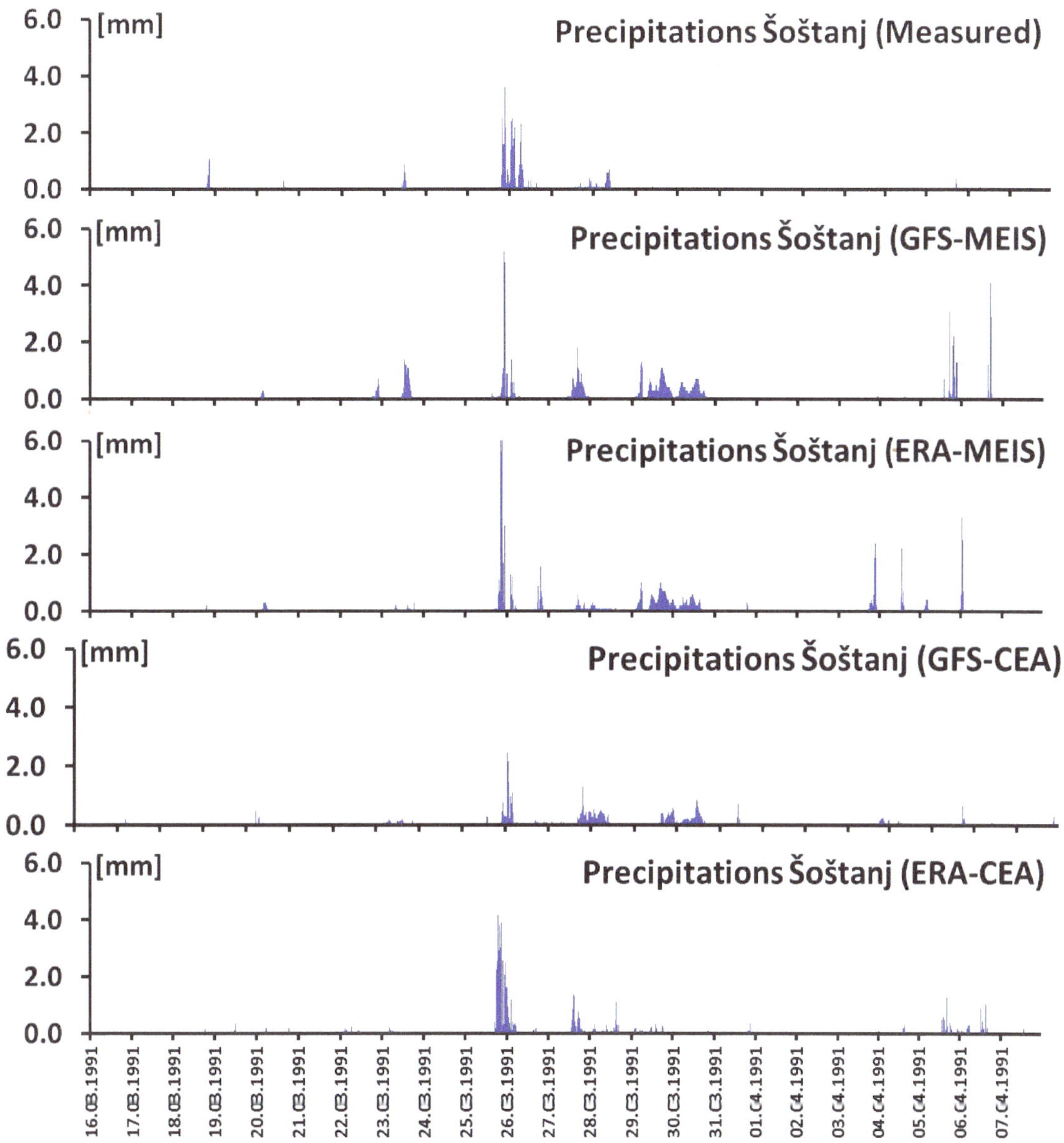

FIG. 58. Precipitation timeline at the Šoštanj location for the measurements (top) and simulations from MEIS-1, MEIS-2, CEA-1 and CEA-2. (Top two graphs and bottom (reproduced from Ref. [15], with permission courtesy of Harmo Initiative).

5.4.3. Global solar radiation

Predicted timelines of global solar radiation matched the measurements reasonably well (Fig. 55). Comparison of the predicted and measured daily cycle indicated that in the high range (500 and 800 W/m^2), the model predictions were slightly too high (Fig. 56). The distribution of errors for the daily cycle (Fig. 57) suggests that errors above 200 W/m^2 were common (especially before noon and between 14:00 and 15:00), although most errors were below 200 W/m^2 (considered an acceptable level) through all time intervals.

5.4.4. Precipitation

Predictions of precipitation at the Šoštanj station were surprisingly good throughout the time period being modelled (Fig. 58). Washout and wet deposition are especially important: heavy rainfall at the time of a release can result in high wet deposition in the vicinity of an emissions source. Therefore, assessment of wet deposition depends on accurate estimates of precipitation.

5.4.5. Total assessment

In general, predicted values for meteorological variables corresponded well with the measurement results at the fine spatial and temporal scales used in this exercise. Improvements are needed for wind speed (which tended to be overpredicted for layers near the ground) and for wind direction. Statistical analysis (Fig. 59) indicates that air temperature is generally well estimated, or occasionally underpredicted, while models tended to overpredict the wind speed.

5.5. CONCLUSIONS

Meteorological forecasts are used in many countries to predict the atmospheric dispersion of radionuclides released from a facility in the event of an accident, and to plan the corresponding emergency actions that would be necessary. Therefore, it is important to evaluate how well the forecasting models predict those variables that most affect the atmospheric dispersion modelling. This exercise analysed results for an area of complex terrain, for which both meteorological modelling and atmospheric dispersion modelling present a major challenge. Results are presented for the specific variables identified; future work could include analysis of atmospheric dispersion predictions based on meteorological forecasts.

5.6. ACKNOWLEDGEMENT

The projects (Method for the Forecasting of Local Radiological Pollution of Atmosphere using Gaussian Process Models, L2-8174, and Modelling the Dynamics of Short-Term Exposure to Radiation, L2-2615) were financially supported by the Slovenian Research Agency.

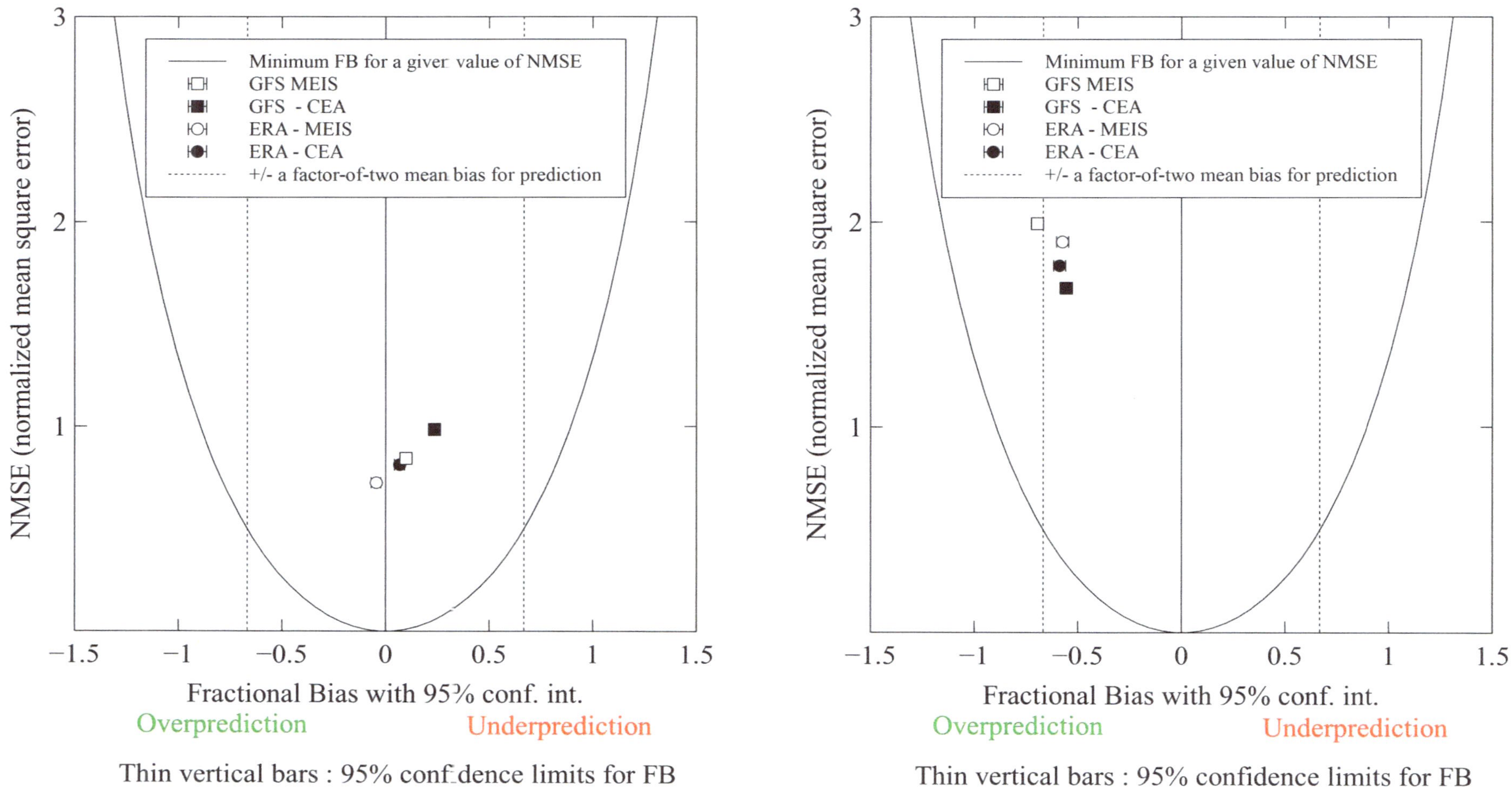

FIG. 59. Statistical analysis of the full data for air temperature (left) and wind speed (right) from 15 March 1991 (07:00) to 5 April 1991 (12:00 (adapted from Ref. [15], with permission courtesy of Harmo Initiative).

6. MID-RANGE ATMOSPHERIC DISPERSION EXERCISE–CHALK RIVER

6.1. INTRODUCTION

The Chalk River exercise was based on monitoring data for routine releases of radioactivity (^{41}Ar) from a research reactor at Chalk River (Canadian Nuclear Laboratories) in Ontario, Canada. The exercise is intended to provide an opportunity to test model predictions for dispersion over a range of a few kilometres from a release location.

Input information for the exercise included a description of the site (geography) and of the source of the releases, source term data, and meteorological data. The primary modelling endpoint for this exercise was the gamma dose rate (μR/h[4]) at a downwind monitoring station for comparison with measurements; the input information could also be used to generate contour maps of ^{41}Ar dose rates at designated times. The overall time period considered was 16–31 July 2008.

6.2. DESCRIPTION OF THE TEST SITE

The Chalk River site consists of gently sloping terrain that has usually been assumed to be flat for modelling purposes. The area is crossed by a wide river that makes the area complex for modelling purposes. There is about a 20 m difference in elevation between the stack (release) location and the monitoring station, over a distance of more than 2 km.

The geographical coordinates (estimated by a Canadian participant of the WG from online sources) are as follows:

Stack: 46.05381 deg N, -77.37605 deg E

Receptor (A117): 46.06283 deg N, -77.40408 deg E

Meteorological station (Perch Lake Tower): 46.03756 deg N, -77.37959 deg E

The receptor site (monitoring station A117) is about 2 km from the stack (WNW 297 deg; wind direction ESE 117 deg). Additional information about the site, including a map, is available in Ref. [29].

6.3. RELEASE SITE AND SOURCE TERM

The source of the releases of ^{41}Ar is the NRU (National Research Universal) research reactor at Chalk River. Releases from this reactor were assumed to be nearly constant unless the reactor was shut down. The stack height is ~46 m (150 ft = 45.72 m), and the stack diameter is 2.0 m. The temperature of the plume was assumed to be constant at 20°C, and the exit velocity was 10 m/s. The emissions data were provided in terms of weekly averages.

The reported releases of ^{41}Ar during the time period of 16 July 2008 to 6 August 2008 were between 2×10^{14} and 3.5×10^{14} Bq/week (provided by a Canadian participant of the WG, based on the annual safety review for 2009).

[4] 1 μR/h = 0.01 μSv/h

6.4. DESCRIPTION OF THE METEOROLOGY

Meteorological data at the Perch Lake Tower and precipitation data (mm) at the Acid Rain Station (A117) for 16–31 July 2008 were provided to participants in electronic form (see the complementary Electronic Appendix to this publication). The following meteorological information was included, at a time interval of 15 mins:

- Temperature (°C; height = 2 m, 30 m, 60 m);
- Wind direction (degrees; height = 30 m);
- Wind speed (km/h; height = 30 m);
- Sigma theta of the wind direction (degrees; height = 30 m).

Times are in in EST (Eastern Standard Time; UTC-5).

6.5. DATA FOR COMPARISON WITH MODEL PREDICTIONS

Monitoring data (gamma dose rate, μR/h, attributed to ^{41}Ar) are available for the time period of 16–31 July 2008 at the Acid Rain Station (A117)[5]. The ARMMS (Ambient Radiation and Metrological Monitoring System) gamma radiation monitor at A117 (μR/h, 10 Hz) is designed to monitor high accidental levels of radiation; the lower threshold reading is set at about 17 μR/h [30]. The monitor uses a Geiger-Müller tube and a 0.1 Hz record. Background gamma radiation measurements with a mobile survey meter were found to be in the range of 0.68 ± 0.08 μR/h [30].

The monitoring data were used as test data and were not provided to participants until after they had submitted their predictions. This permitted blind testing of the model predictions.

6.6. MODELLING ENDPOINTS

Participants in the exercise were asked to provide predictions for the time series of dose rates from ^{41}Ar (μR/h) at the monitoring station A117 for comparison with measurements. It was desirable to predict values for every 15 mins. This modelling endpoint is the subject of most of the discussion that follows.

Predictions for the ^{41}Ar activity concentrations in air were not specifically requested but were reported by one participant.

Contour maps of dose rate at specific times were suggested as an optional endpoint for the exercise and were submitted by two participants.

The analysis of the model predictions for the exercise included calculation of time integrated endpoints. Specifically, the time integrated dose rate was calculated for each submitted time series, in effect, the dose to a hypothetical receptor located at the A117 monitoring station.

---

[5] Also included in the complementary Electronic Appendix to this publication.

6.7. MODELS USED IN THE EXERCISE

Five models were used in the exercise (Table 11), i.e. ARTM [20] and LASAIR from Germany, JRODOS (with LASAT and with RIMPUFF) from Switzerland, and RG 1.145 [31] and SPRAY from Slovenia [32]. Table 11 provides detailed descriptions of the models and modelling approaches, together with key assumptions and parameter values used in the modelling for the Chalk River exercise.

The RG 1.145 model is a Gaussian plume model, designed for emergency response. ARTM, SPRAY, LASAT (used in JRODOS) and LASAIR are Lagrangian particle models. RIMPUFF (used in one simulation with JRODOS) is a Lagrangian puff diffusion model. SPRAY is intended for universal use. The two German models were designed for different purposes, i.e. LASAIR is essentially for emergency response in urban areas, while ARTM is for long term dispersion calculations for annual reporting purposes and mostly simulates rural areas. LASAT and RIMPUFF are also designed for emergency purposes. More information about individual models as used in this exercise is provided in Appendix I.

For most of the models, several simulation runs were carried out, with some differences in the input parameters. RG 1.145, the only Gaussian model in the comparison, was used with two different methods to determine the stability class (delta T and sigma theta). SPRAY simulations were performed based both on measured data from the Perch Lake Tower (diagnostic approach), and on meteorological forecast data obtained using the WRF (Weather Research and Forecasting) model (prognostic approach). ARTM was run with and without consideration of plume rise. JRODOS was used for five simulations (Table 12) with two different domain sizes and two different time resolutions (calculation time steps); four simulations used LASAT and one used RIMPUFF. One JRODOS simulation with LASAT used meteorological forecast data (MARS-LASAT); the rest of the JRODOS simulations used the measured data.

Differences between the models included the domain size, which ranged from 12 km^2 to 250 km^2, and the grid sizes, which ranged from 5 m to 400 m. SPRAY, ARTM, and JRODOS integrated topography data into the simulations. The time resolution (calculation time step) of the simulations ranged from 15 minutes to 60 minutes.

6.8. ANALYSIS OF MODELLING RESULTS

Most of the discussion and analysis of modelling results pertain to predicted time series of ^{41}Ar dose rates (μR/h). Additional endpoints discussed in this section include predictions of activity concentrations of ^{41}Ar in air (Bq/m^3), time integrated endpoints, and contour maps of ^{41}Ar dose rates.

6.8.1. Time series of ^{41}Ar dose rates (μR/h)

Examples of predicted time series for the whole 16 day time period are provided in Figs 60–69. For ease in making comparisons among model predictions or between model predictions and measurements, several shorter (2 day) sections of the time series were examined in more detail. These included 17–18 July 2008 (midnight to midnight), 25–26 July 2008 (midnight to midnight), 27–28 July 2008 (midnight to midnight), and 29–31 July 2008 (noon to noon). These time periods generally had the highest measured or predicted dose rates.

TABLE 11. SUMMARY OF MODELS AND MODELLING APPROACHES USED FOR THE CHALK RIVER ^{41}AR EXERCISE

Model name	**SPRAY (diagnostic) 'lgm01'**	**SPRAY (prognostic) 'lgm02'**	**RG 1.145 (delta T) 'RG 1.145 – DT58'**	**RG 1.145 (sigma theta) 'RG 1.145 – SigTh'**
Participant and country	MEIS d.o.o., Slovenia	MEIS d.o.o., Slovenia	MEIS d.o.o., Slovenia	MEIS d.o.o., Slovenia
Purpose of model	Universal	Universal	Emergency response	Emergency response
Type of model	Diagnostic wind field model, Lagrangian particle model	Diagnostic wind field model, Lagrangian particle model	Gaussian model	Gaussian model
Number of Lagrangian particles	40 000 per 30 minutes	40 000 per 30 minutes	Not applicable	Not applicable
Domain size/calculation range	5 km × 5 km	5 km × 5 km	10 km × 10 km	10 km × 10 km
Grid size	50 m	50 m	100 m	100 m
Grid height	20 levels up to 3000 m above the ground, lowest level is 10 m tall	20 levels up to 3000 m above the ground, lowest level is 10 m tall	Not applicable	Not applicable
Release height	46 m + plume rise	46 m + plume rise	43 m + Holland equation	43 m + Holland equation
Receptor height	10 m average	10 m average	0 m	0 m
Stability class	Measured temperature lapse rate and wind speed	Forecasted temperature lapse rate, wind speed and solar radiation	Measured temperature lapse rate	Measured wind fluctuations
Wind speed and direction	Measured 30 minute average	Forecasted 30 minute average vertical profile	Measured 15 min	Measured 15 min
Air temperature	Measured 30 minute average	Forecasted 30 minute average vertical profile	Not applicable	Not applicable
Dispersion parameters	Calculated every 30 minutes	Calculated every 30 minutes	Stability class calculated every 15 minutes	Stability class calculated every 15 minutes
Plume rise	Calculated	Calculated	Holland equation	Holland equation
Depletion	Not considered	Not considered	Not considered	Not considered
Release time step	Constant	Constant	Constant	Constant
Calculation time step	30 minutes	30 minutes	15 minutes	15 minutes
Simulation time	16 days	16 days	16 days	16 days
Terrain/topography	Extracted from Canadian Digital Elevation Model, 50 m resolution	Extracted from Canadian Digital Elevation Model, 50 m resolution	Not considered	Not considered
Friction coefficient or friction velocity	z_0 industrial = 1.0 m z_0 mixed forest = 1.3 m z_0 water = 0.01 m	z_0 industrial = 1.0 m z_0 mixed forest = 1.3 m z_0 water = 0.01 m	Not considered	Not considered
Time to set up and run	Set up: 3 days Run: 6 hours	Set up: 3 days Run: 6 hours	Set up: 3 days Run: 5 minutes	Set up: 3 days Run: 5 minutes
Time to process results	1 day	1 day	1 day	1 day
Total source term	6×10^{14} Bq	6×10^{14} Bq	6×10^{14} Bq	6×10^{14} Bq
Gamma cloud shine	Simple cloud shine routine ^{41}Ar = 2.20×10^{-10} Sv/h per Bq/m^3	Simple cloud shine routine ^{41}Ar = 2.20×10^{-10} Sv/h per Bq/m^3	^{41}Ar = 2.20×10^{-10} Sv/h per Bq/m^3	^{41}Ar = 2.20×10^{-10} Sv/h per Bq/m^3

TABLE 11. (cont.)

Model name	**ARTM v. 2.8**	**LASAIR v. 5**	**JRODOS**
Participant and country	Margit Pattantyús-Ábrahám, Germany	Hartmut Walter, Germany	Lucia Federspiel, Switzerland
Purpose of model	Atmospheric dispersion model for regulatory purposes, dealing with planned releases	Emergency response	Emergency response
Type of model	Diagnostic wind field model, Lagrangian particle model	Lagrangian particle model	Lagrangian particle model LASAT
Number of Lagrangian particles	6.3×10^7	6.0×10^7	2.21×10^7
Domain size/calculation range	3800 m × 3250 m	40 km × 40 km	(2 × 38.4 km) × (2 × 38.4 km) (2 × 4.8 km) × (2 × 4.8 km)
Grid size	25 m	5 m, 10 m, 20 m, 40 m, 80 m, 160 m, 320 m	4 grids: 50 m, 100 m, 200 m, 400 m
Grid height	19 levels up to 1500 m. Lowest level is 3 m height.	19 levels up to 1500 m. Lowest level is 3 m height.	11 vertical levels up to 1000 m. Lowest level is 1 m height.
Release height	50 m	50 m (46 m + approximately 4 m plume rise)	46 m + plume rise
Receptor height	1.5 m	1.5 m	1 m
Stability class	According to weather, calculated from temperature lapse rate	According to weather, calculated from temperature lapse rate	According to weather, calculated from temperature lapse rate
Wind speed and direction	According to meteorological measurements, averaged for hour	According to meteorological measurements each 15 minutes	According to meteorological measurements, averaged for hour
Air temperature	Not considered in ARTM	Not considered in LASAIR	According to meteorological measurements, averaged for hour
Dispersion parameters	Optimized for an hour, according to the stability class	Optimized for each quarter of an hour, according to the stability class. Turbulence scheme according to German guideline VDI 3783 Blatt 8	Optimized for an hour, according to the stability class
Plume rise	Considered and not considered	Considered	Considered
Depletion	Not considered	Considered in LASAIR V5	Considered
Release time step	Weekly constant	Weekly constant	Weekly constant
Calculation time step	1 hour	15 minutes	1 hour
Simulation time	3 weeks	16 days (385 h)	16 days (384 h)
Terrain/topography	Extracted from Canadian Digital Elevation Model, 20 m resolution	No orographical data in this area available for LASAIR	Elevation: Extracted from SRTM Digital Elevation database by NASA, 90 m resolution Land use: from Canadian Digital Elevation Model, 50 m resolution

TABLE 11. (cont.)

Model name	ARTM v. 2.8	LASAIR v. 5	JRODOS
Friction coefficient or friction velocity	Roughness length z_0 = 1.5 m	Roughness length z_0 = 1.5 m Roughness length water = 0.01 m Roughness length village = 1.0 m	Roughness length z_0 = 0.1 m (default) z_0 industrial = 1.0 m z_0 mixed forest = 1.3 m z_0 water = 0.01 m
Time to set up and run	Getting topography data: 6 hrs. Set up: 4 days (including non-successful runs). Run: 30 min	Set up: 2 days (including test runs). Run: 7 h (depending on performance of PC)	Getting topography and land use data: 1 day. Getting weather data: 2 days. Set up: 4 weeks (incl. resolution of several calculation problems (implementation ^{41}Ar and Chalk River site, environmental data for Canada, LASAT >10days, test runs), Run: ~5h
Time to process results	3 days	1 day	10 days
Total source term	5×10^{14} Bq	6.2×10^{14} Bq (average release of each week)	6.47×10^{14} Bq
Gamma cloud shine	Gamma submersion, finite volume source	Effective Gamma radiation dose rate for adults, ground and cloud shine	Cloud effective gamma dose rate for adults (ground effective gamma dose = 0 for noble gases) $^{41}Ar = 2.23 \times 10^{-1}$ nSv/h per Bq/m^3

Abbreviations:
ARTM = Atmospheric Radionuclide Transport Model (http://www.bfs.de/EN/topics/ion/environment/air-soil/emission-monitoring/artm.html).
JRODOS = Java-based version of RODOS (Real time Online Decision Support system) (https://resy5.iket.kit.edu/JRODOS/).
LASAIR = Lagrangian Simulation of the dispersion (Ausbreitung) and Inhalation of Radionuclides.
LASAT = Lagrangian Simulation of Aerosol Transport (Lagrangian particle model within JRODOS).
RG 1.145 = model based on USNRC Regulatory Guide 1.145.

TABLE 12. SUMMARY OF SIMULATIONS PERFORMED WITH JRODOS

Simulation name	LASAT 1 h 38.4 km	LASAT 1 h 5 km	LASAT 15 min 38.4 km	RIMPUFF 1 h 38.4 km	LASAT MARS 38.4 km
Weather data	Met station, 1 h average	Met station, 1 h average	Met station, 15 min	Met station, 1 h average	ECMWF-MARS archive[a]
Atmospheric dispersion model	LASAT	LASAT	LASAT	RIMPUFF	LASAT
Plume rise	Yes	Yes	Yes	Yes	Yes
Number of grids (rings)	4	1	4	4	4
Spatial resolution, 1° grid	50 m	50 m	50 m	50 m	50 m
Calculation domain	(2 × 38.4 km) × (2 × 38.4 km)	(2 × 4.8 km) × (2 × 4.8 km)	(2 × 38.4 km) × (2 × 38.4 km)	(2 × 38.4 km) × (2 × 38.4 km)	(2 × 38.4 km) × (2 × 38.4 km)

[a] ECMWF-MARS = European Centre for Medium-Range Weather Forecasts-Meteorological Archival and Retrieval System.

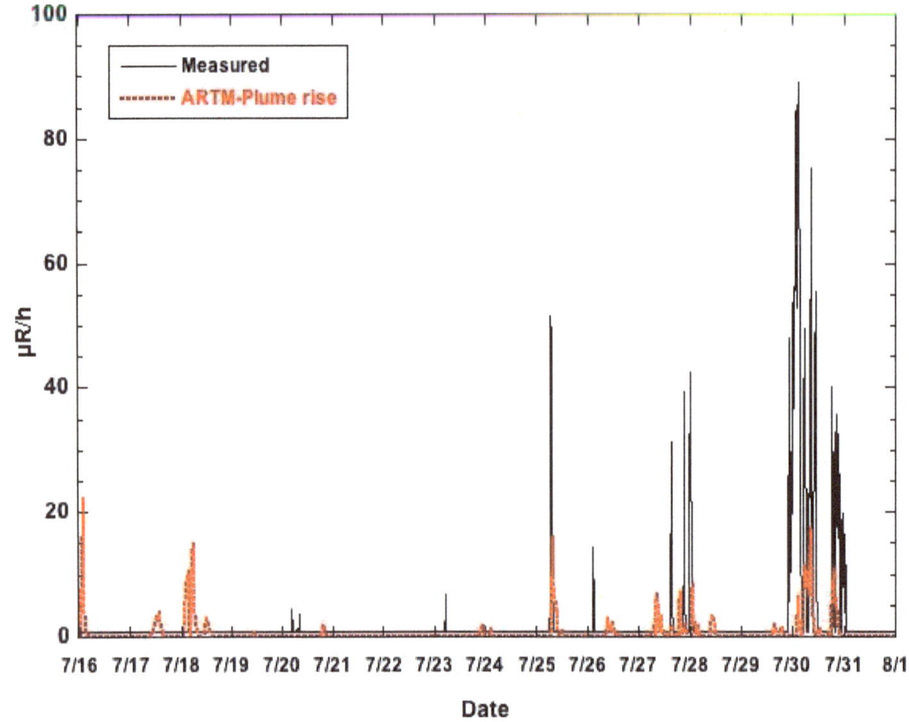

FIG. 60. Comparison of model predictions using ARTM (including plume rise) with measurements for the dose rate (μR/h) from ^{41}Ar at station A117.

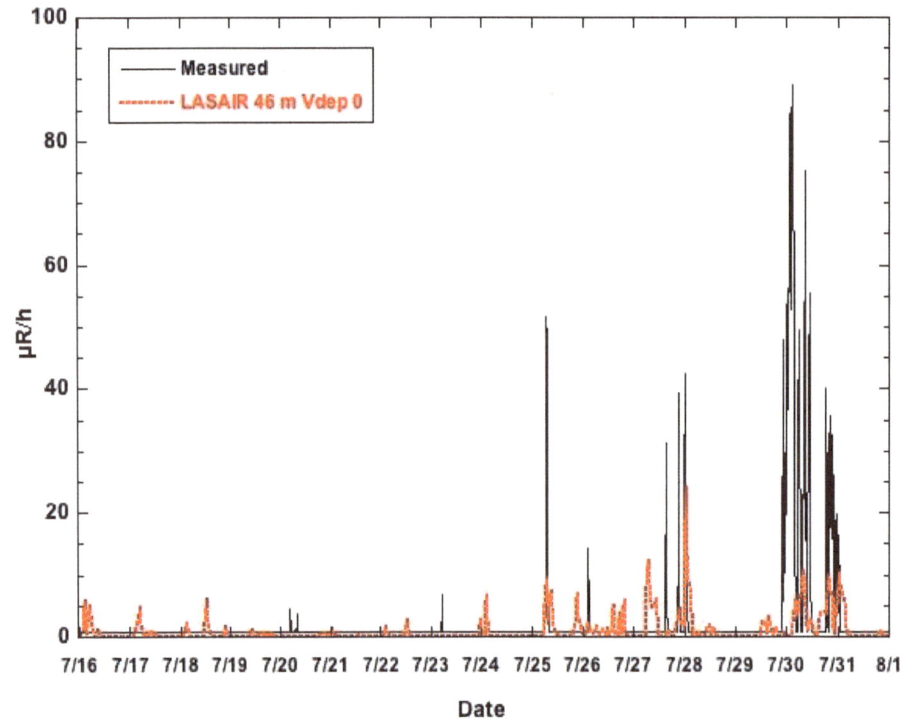

FIG. 61. Comparison of model predictions using LASAIR (release height 46 m; deposition velocity 0) with measurements for the dose rate (μR/h) from ^{41}Ar at station A117.

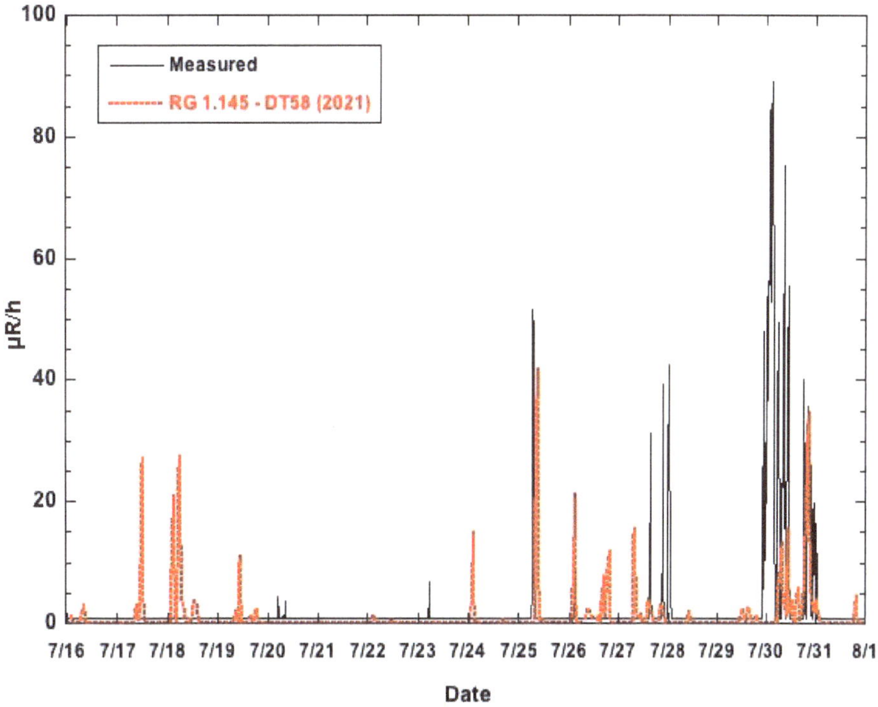

FIG. 62. *Comparison of revised model predictions using RG 1.145 (stability class from delta T, 60 m and 2 m; hourly averages) with measurements for the dose rate (μR/h) from ^{41}Ar at station A117.*

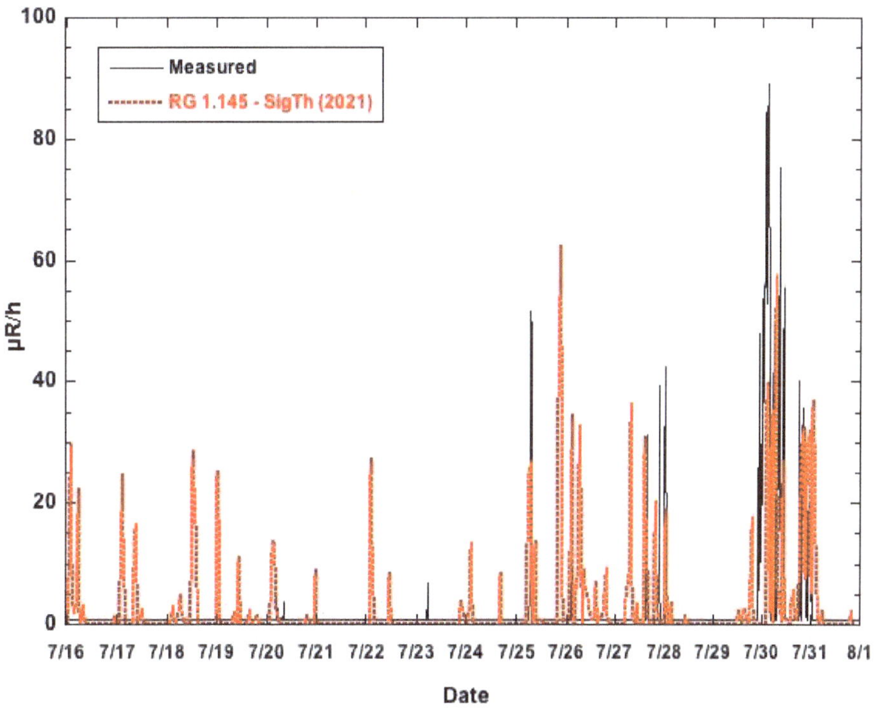

FIG. 63. *Comparison of revised model predictions using RG 1.145 (stability class from sigma theta; hourly averages) with measurements for the dose rate (μR/h) from ^{41}Ar at station A117.*

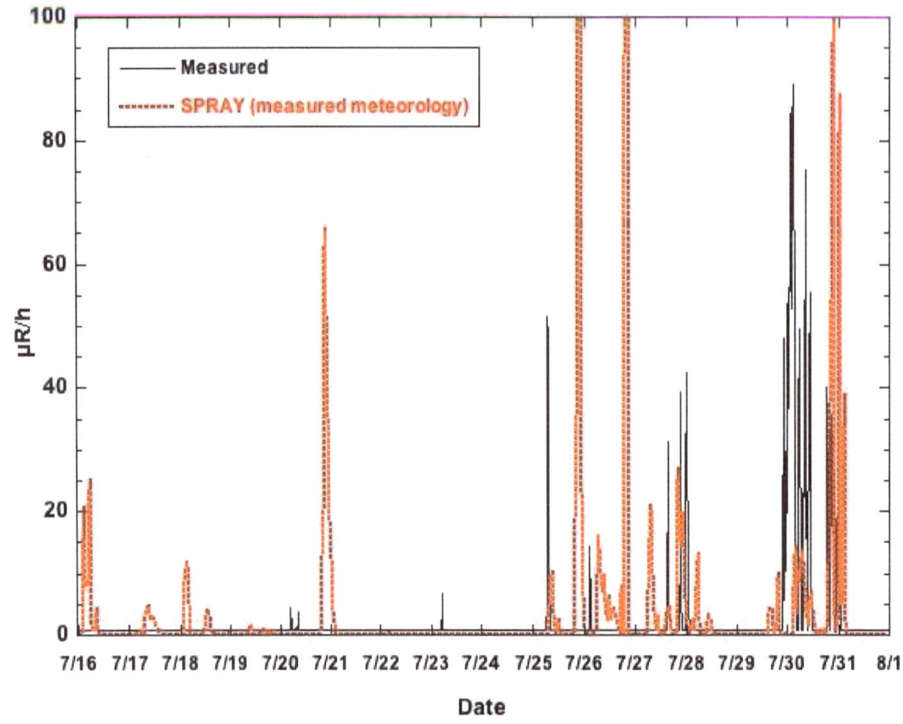

FIG. 64. Comparison of initial model predictions using SPRAY (measured meteorological data) with measurements for the dose rate (μR/h) from ^{41}Ar at station A117.

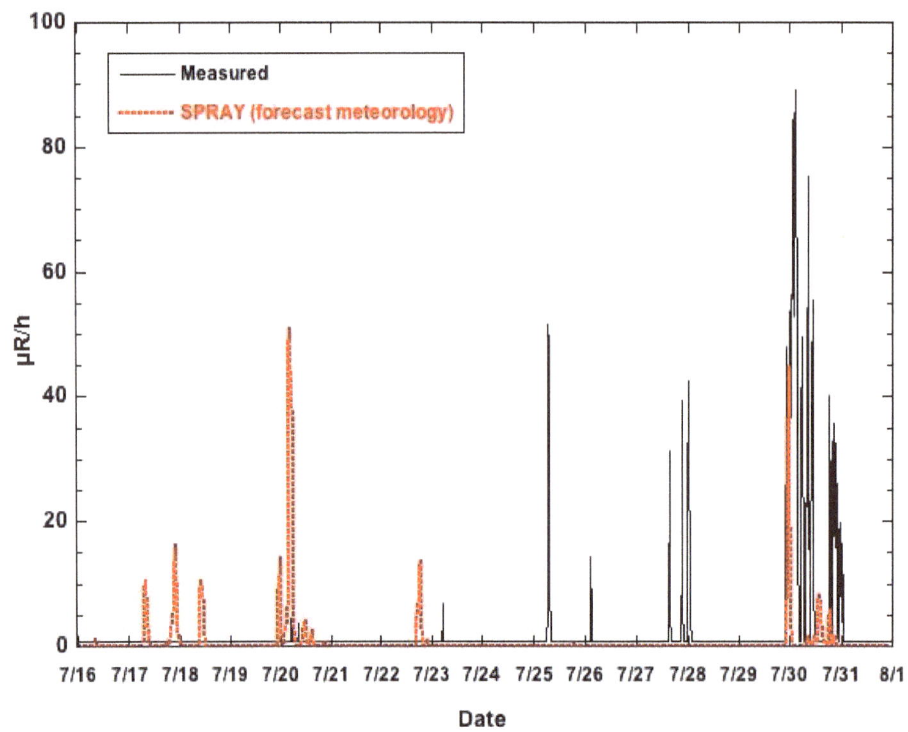

FIG. 65. Comparison of initial model predictions using SPRAY (forecast meteorology) with measurements for the dose rate (μR/h) from ^{41}Ar at station A117.

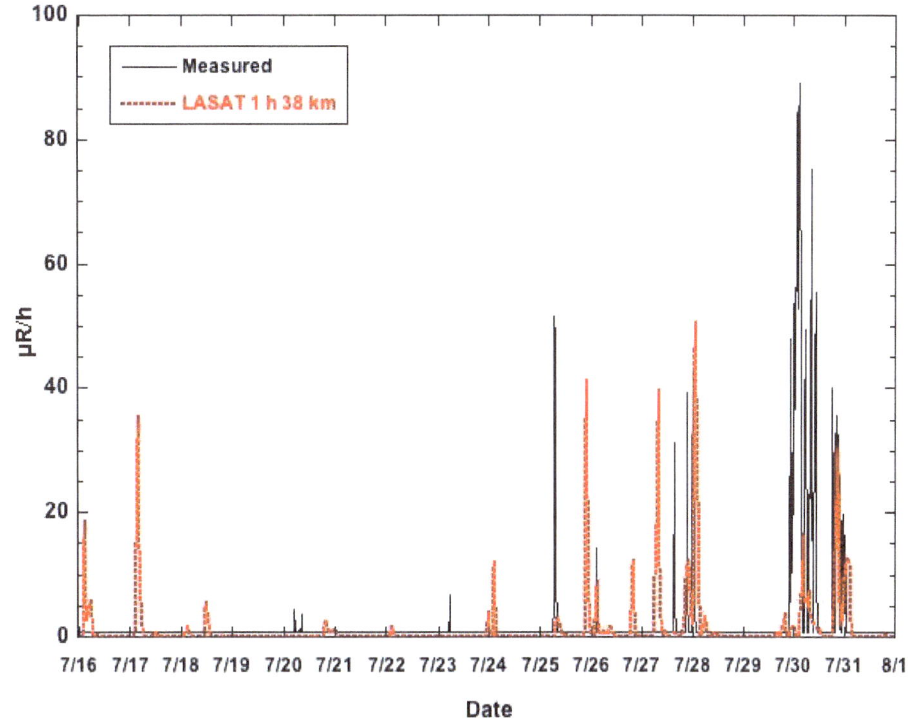

FIG. 66. Comparison of model predictions from JRODOS using LASAT (1 h time step, 38 km grid radius) with measurements for the dose rate (μR/h) from ^{41}Ar at station A117.

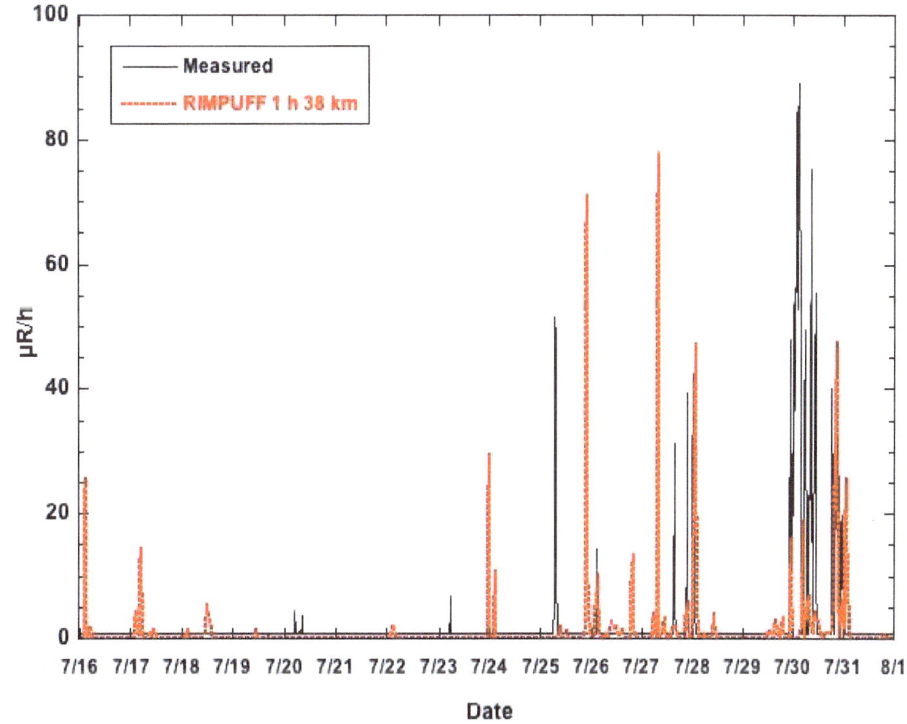

FIG. 67. Comparison of model predictions from JRODOS using RIMPUFF (1 h time step, 38 km grid radius) with measurements for the dose rate (μR/h) from ^{41}Ar at station A117.

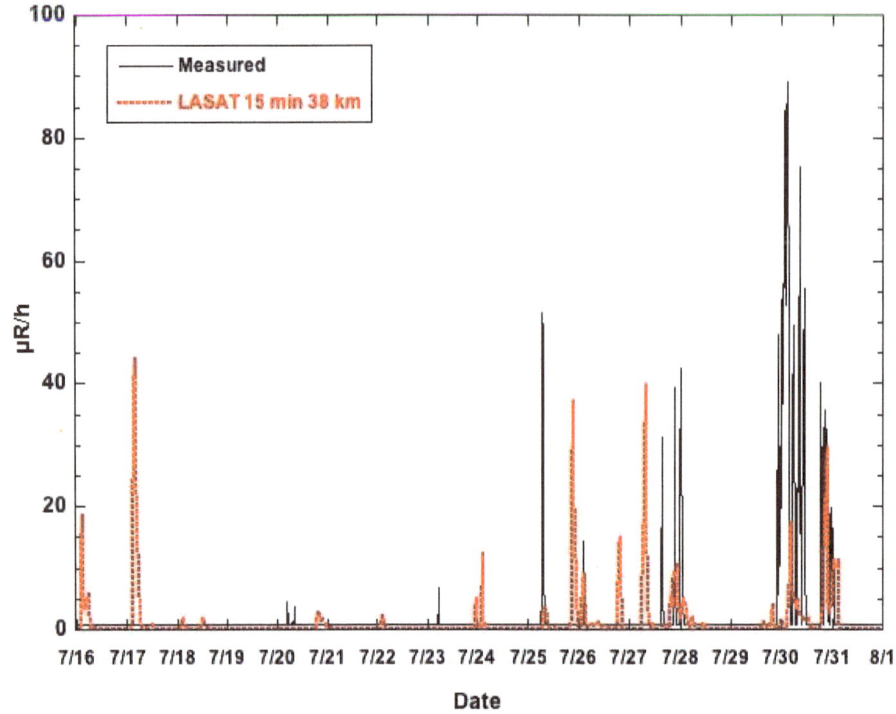

FIG. 68. Comparison of model predictions from JRODOS using LASAT (15 min time step, 38 km grid radius) with measurements for the dose rate (µR/h) from ^{41}Ar at station A117.

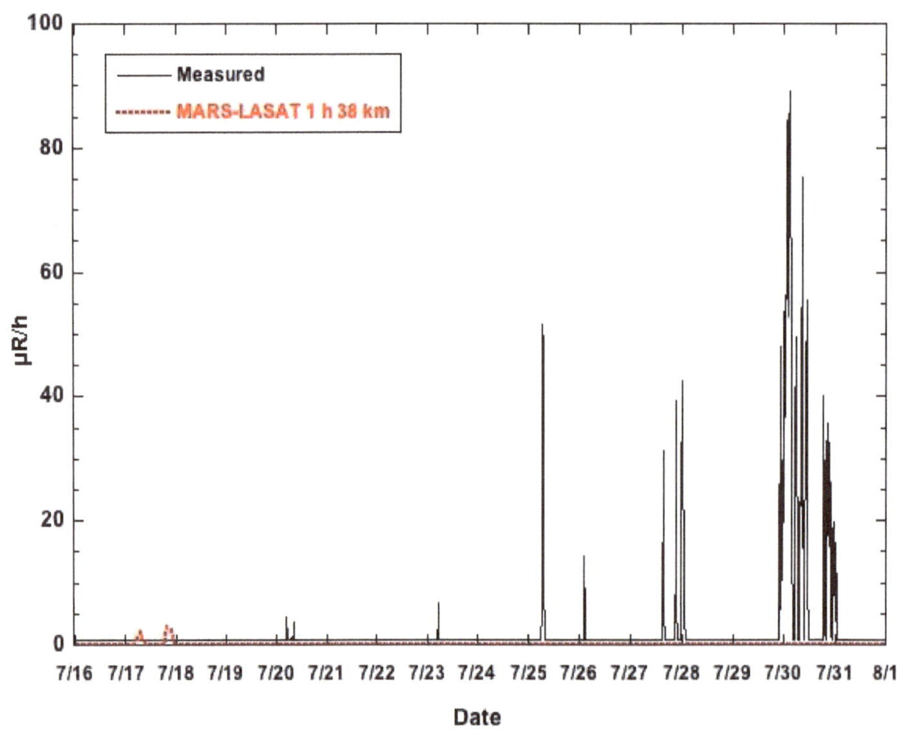

FIG. 69. Comparison of model predictions from JRODOS using MARS forecast meteorology and LASAT (1 h time step, 38 km grid radius) with measurements for the dose rate (µR/h) from ^{41}Ar at station A117.

6.8.1.1. *ARTM model*

Results from the ARTM model for the selected time periods are shown in Figs 70–73. The ARTM model was used for two simulations, with and without inclusion of plume rise. Inclusion of plume rise was associated with greater dispersion and lower predicted dose rates at the receptor location. In most cases, predicted peaks for the two simulations coincided in time, with the 'no plume rise' results giving higher dose rates at any given time point. ARTM predicted increases in dose rate during 17–18 July 2008 that were not seen in the measurements (Figs 60 and 70), as did other models used in the exercise (described later). During the other three time periods, ARTM predicted some of the peaks reasonably well, with the 'no plume rise' version approximating the peak values of the measured dose rate better than the 'plume rise' version. Some measured peaks, e.g. ~16:00 on 27 July 2008 (Fig. 72), ~20:00–04:00 of 29–30 July 2008 (Fig. 73) were not predicted by ARTM, while ARTM predicted some peaks (e.g. ~08:00 on 27 July 2008, Fig. 72) that were not observed in the measurements.

6.8.1.2. *LASAIR model*

Results from the LASAIR model for the selected time periods are shown in Figs 74–77. LASAIR was initially run using a value of 0.01 m/s for the deposition velocity (V_{dep}, including both wet and dry deposition). Noble gases are usually assumed not to be deposited, so a second simulation used a value of 0 m/s for V_{dep}. The assumption of no deposition (V_{dep} = 0 m/s) led to higher values for the predicted dose rate, by factors of 1–19. Very little difference in the two sets of results can be seen in Figs 74–77.

Most peak values of the dose rate predicted with LASAIR are lower than those from other models or those seen in the measurements. Several of the predicted peaks coincided with the measured peaks (e.g. just before 08:00 on 25 July 2008 (Fig. 75)), or the last peak on 28 July 2008 (Fig. 76), or else some of the peaks on 29–30 July 2008 (Fig. 77). As with other models, LASAIR predicted some peak dose rates that were not observed in the measurements (Fig. 61), e.g. ~08:00 on 27 July 2008 (Fig. 76) and small peaks during 17–18 July 2008 (Fig. 74).

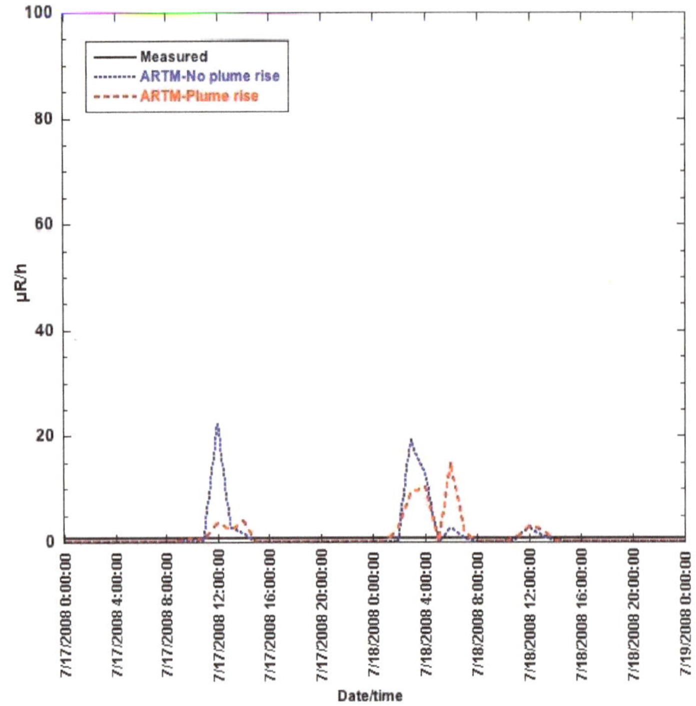

FIG. 70. Comparison of model predictions using ARTM (with and without including plume rise) with measurements for the dose rate (µR/h) from ^{41}Ar at station A117, for 17–18 July 2008 (midnight to midnight).

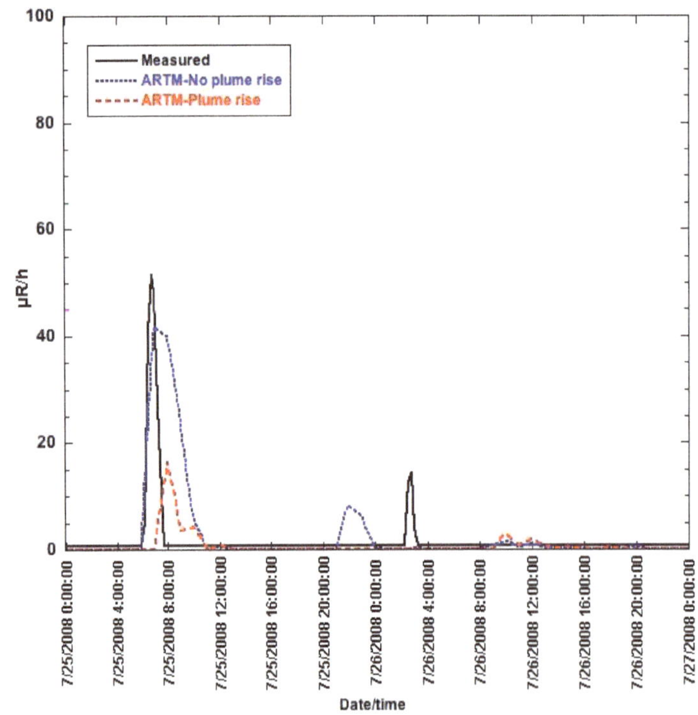

FIG. 71. Comparison of model predictions using ARTM (with and without including plume rise) with measurements for the dose rate (µR/h) from ^{41}Ar at station A117, for 25–26 July 2008 (midnight to midnight).

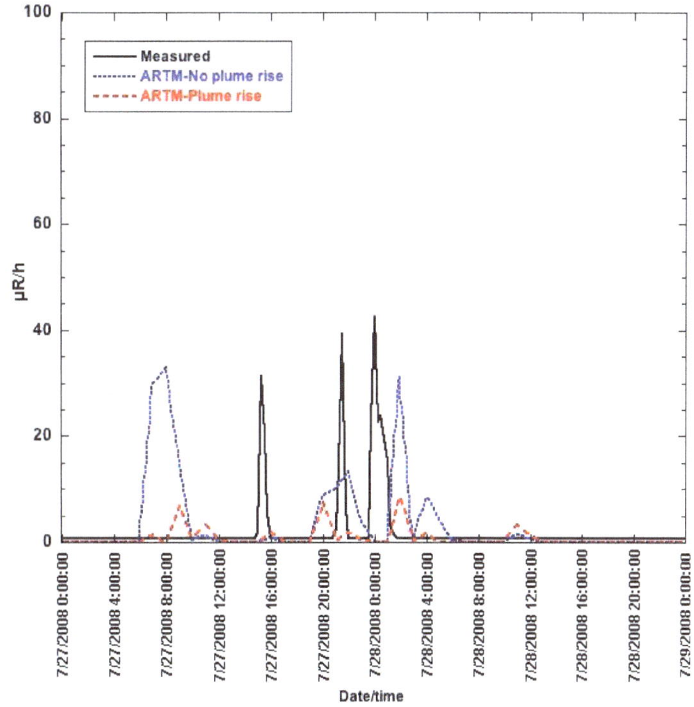

FIG. 72. Comparison of model predictions using ARTM (with and without including plume rise) with measurements for the dose rate (μR/h) from ^{41}Ar at station A117, for 27–28 July 2008 (midnight to midnight).

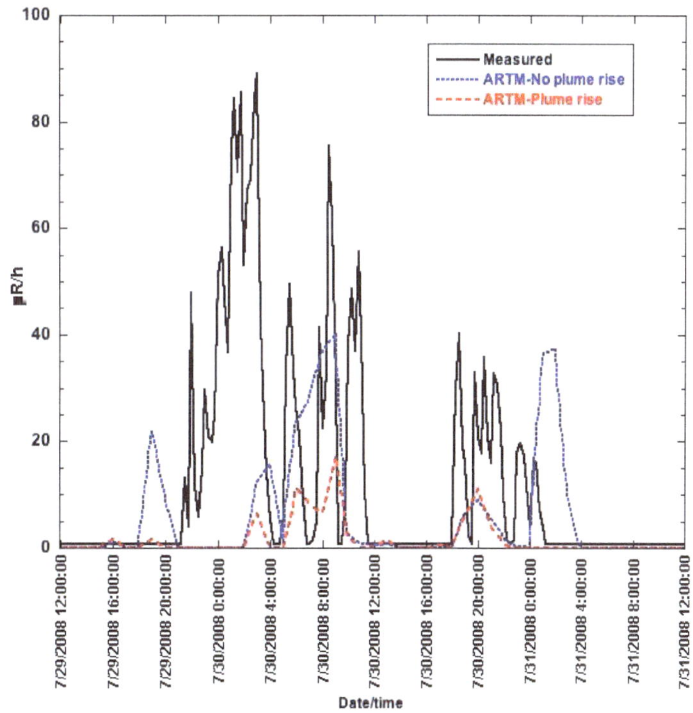

FIG. 73. Comparison of model predictions using ARTM (with and without including plume rise) with measurements for the dose rate (μR/h) from ^{41}Ar at station A117, for 29–31 July 2008 (noon to noon).

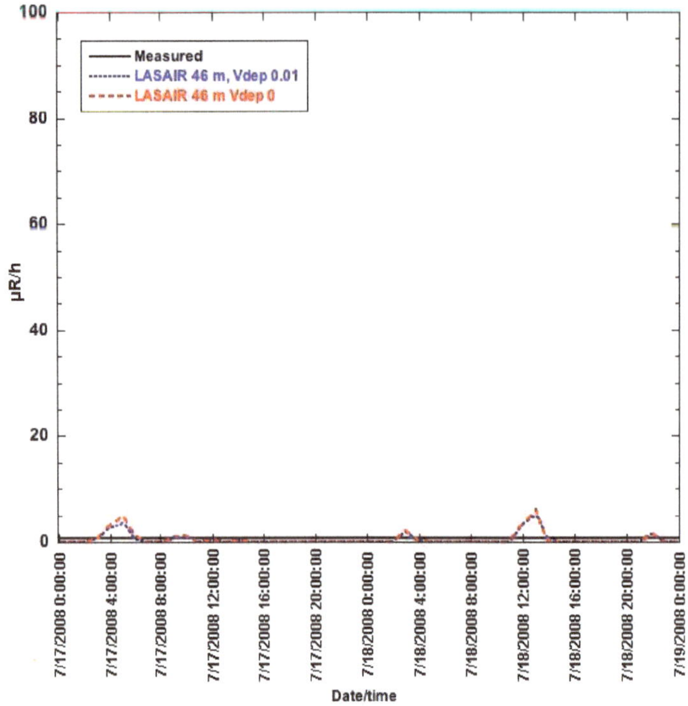

FIG. 74. Comparison of model predictions using LASAIR (with deposition velocity = 0.01 or 0) with measurements for the dose rate (μR/h) from ^{41}Ar at station A117, for 17–18 July 2008 (midnight to midnight).

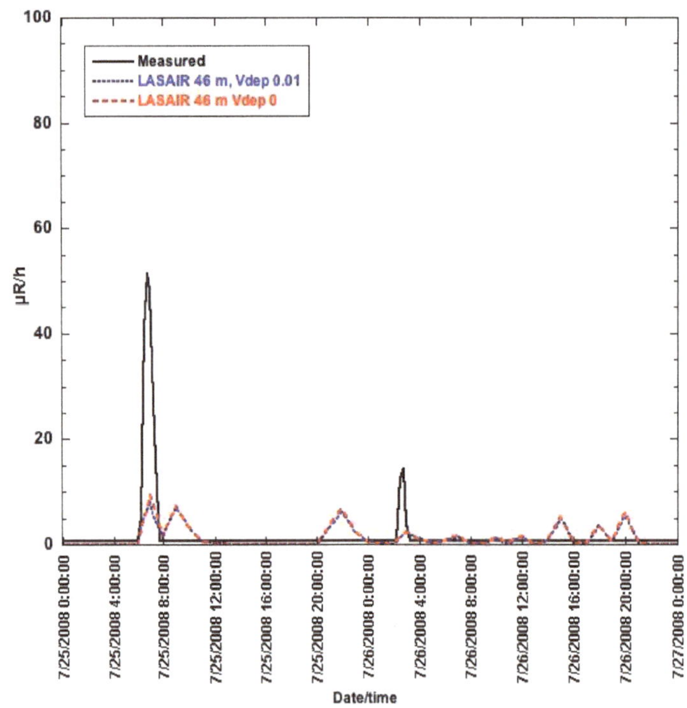

FIG. 75. Comparison of model predictions using LASAIR (with deposition velocity = 0.01 or 0) with measurements for the dose rate (μR/h) from ^{41}Ar at station A117, for 25–26 July 2008 (midnight to midnight).

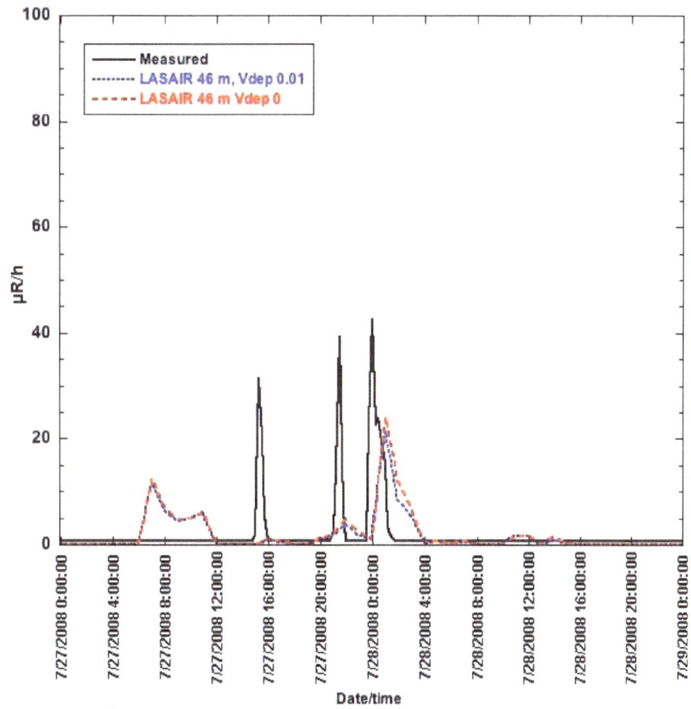

FIG. 76. Comparison of model predictions using LASAIR (with deposition velocity = 0.01 or 0) with measurements for the dose rate (µR/h) from ^{41}Ar at station A117, for 27–28 July 2008 (midnight to midnight).

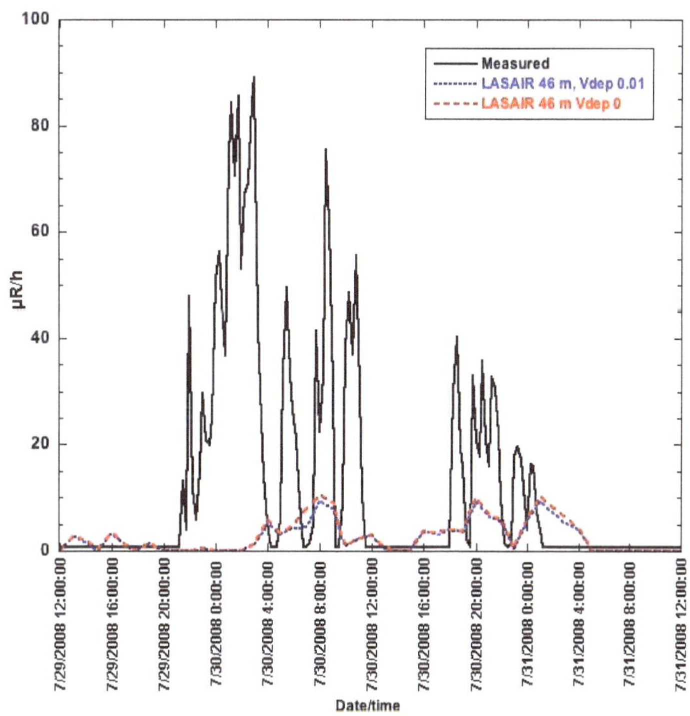

FIG. 77. Comparison of model predictions using LASAIR (with deposition velocity = 0.01 or 0) with measurements for the dose rate (µR/h) from ^{41}Ar at station A117, for 29–31 July 2008 (noon to noon).

6.8.1.3. RG 1.145 model

Results from four simulations with the RG 1.1.45 model for the selected time periods are shown in Figs 78–85. The RG 1.145 model was the only Gaussian diffusion model used in this exercise. It was run with two different approaches for estimating the stability class. The first, labelled DT58 in the figures, used delta T (the difference in the temperatures) between the 60 m and 2 m measurements (hence DT58, for delta T for the 58 m difference in height of the measurements). The other approach, labelled SigTh in the figures, used the standard deviation of the wind direction (sigma for the angle theta) to estimate the stability class. An initial set of model predictions using both approaches was submitted early in the exercise (Figs 78, 80, 82 and 84); these predictions were reported for 15 minute time intervals (corresponding to the meteorological measurements). A second set of model predictions consisted of hourly averages of the 15 minute predictions from the first set of model predictions (Figs 79, 81, 83 and 85). As shown in the figures, the revised predictions generally consisted of shorter, broader peaks.

As with the other models, in several cases the predicted peaks coincided with the measured peaks, e.g. ~08:00 on 25 July 2008 (Figs 80 and 81), the second and fourth peaks on 27–28 July 2008 (Figs 82 and 83), and some of the peaks on 29–30 July 2008 (Figs 84 and 85). Again, as with other models, RG 1.145 predicted peak dose rates that were not observed in the measurements, e.g. 17–18 July 2008 (Figs 78 and 79), between 20:00 and 00:00 on 25 July 2008 (Fig. 81), and around 08:00 on 27 July 2008 (Figs 82 and 83).

Looking primarily at the revised predictions (Figs 62. 63, 79, 81, 83 and 85), the sigma theta approach seemed to yield predictions more nearly resembling the measurements in timing and height of the peaks. Both approaches led to predicted peaks that were not seen in the measurements, with the sigma theta approach producing more of those.

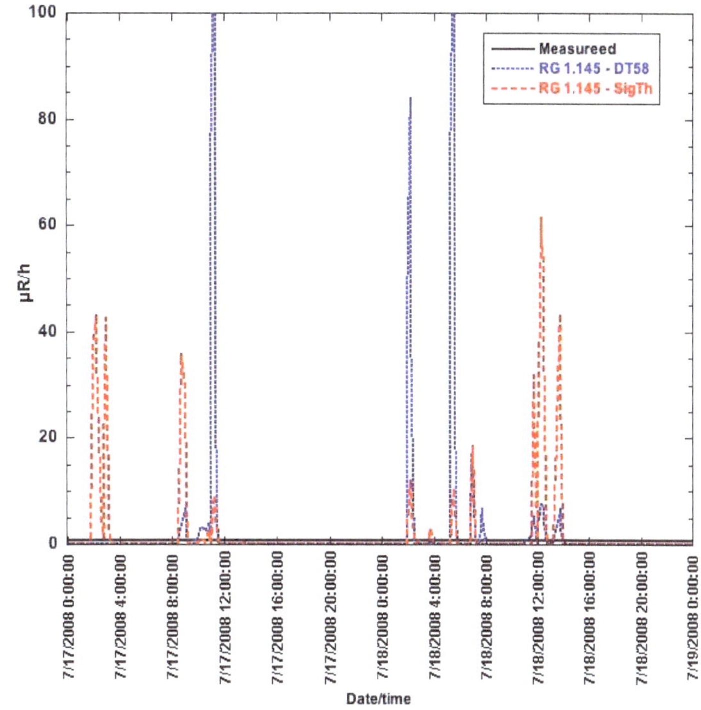

FIG. 78. Comparison of initial model predictions using RG 1.145 (with stability class calculated from either delta T or sigma theta) with measurements for the dose rate (μR/h) from ^{41}Ar at station A117, for 17–18 July 2008 (midnight to midnight).

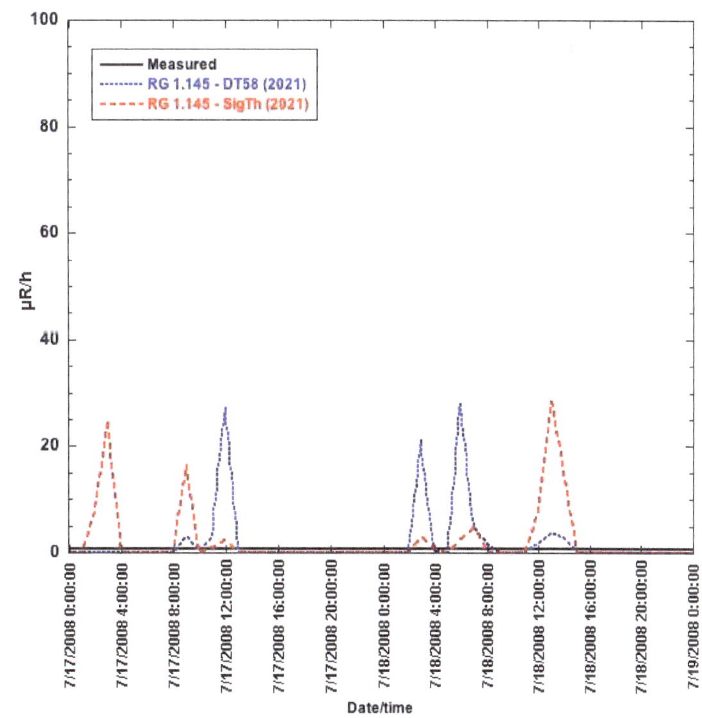

FIG. 79. Comparison of revised model predictions using RG 1.145 (with stability class calculated from either delta T or sigma theta) with measurements for the dose rate (μR/h) from ^{41}Ar at station A117, for 17–18 July 2008 (midnight to midnight).

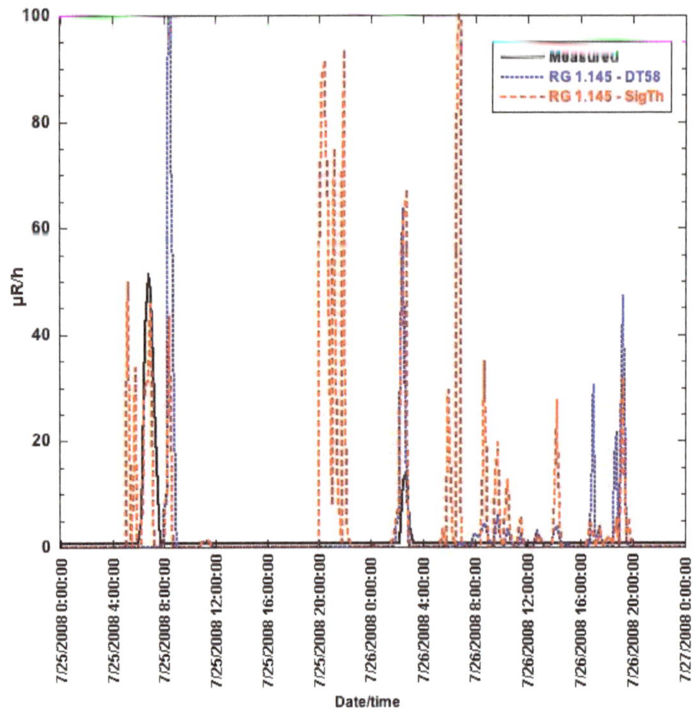

FIG. 80. Comparison of initial model predictions using RG 1.145 (with stability class calculated from either delta T or sigma theta) with measurements for the dose rate (μR/h) from ^{41}Ar at station A117, for 25–26 July 2008 (midnight to midnight).

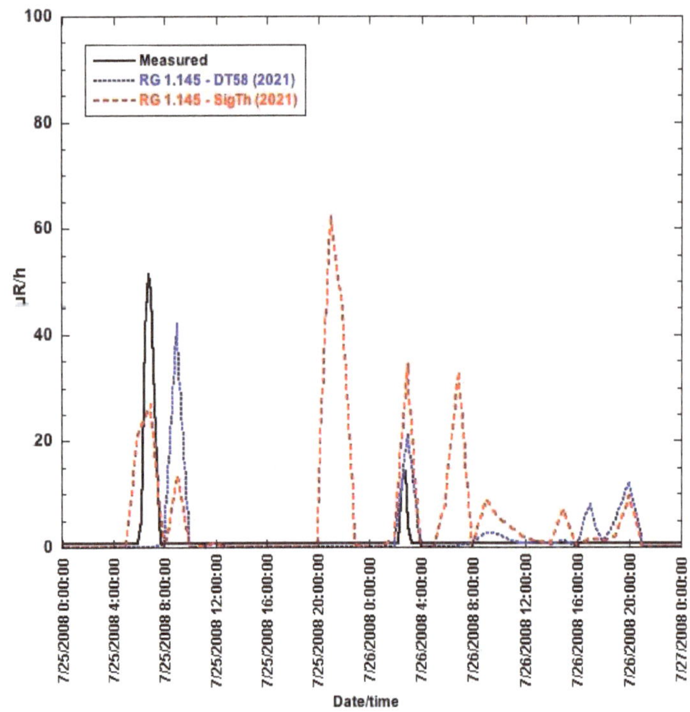

FIG. 81. Comparison of revised model predictions using RG 1.145 (with stability class calculated from either delta T or sigma theta) with measurements for the dose rate (μR/h) from ^{41}Ar at station A117, for 25–26 July 2008 (midnight to midnight).

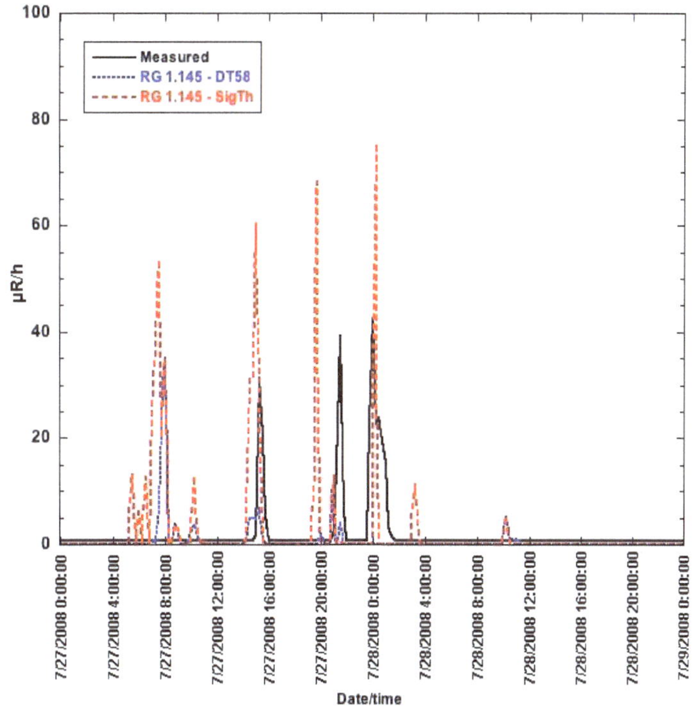

FIG. 82. Comparison of initial model predictions using RG 1.145 (with stability class calculated from either delta T or sigma theta) with measurements for the dose rate (μR/h) from ^{41}Ar at station A117, for 27–28 July 2008 (midnight to midnight).

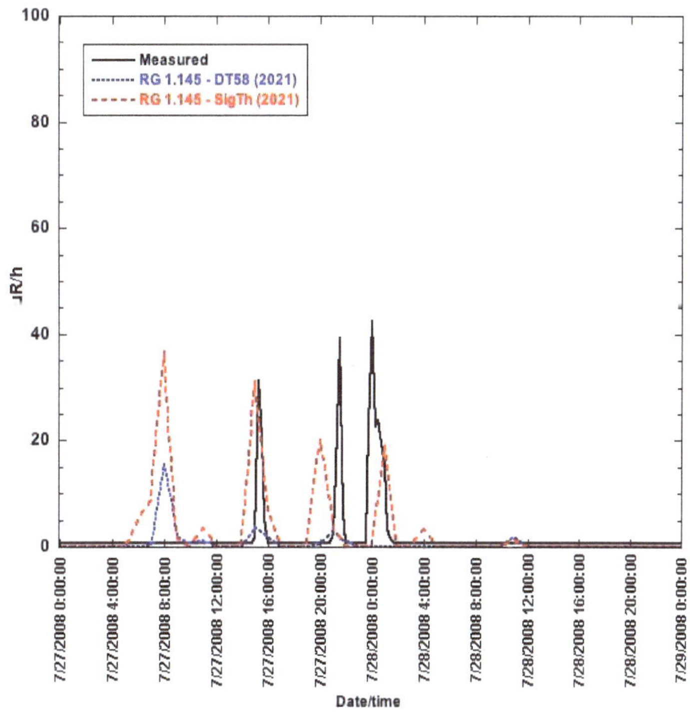

FIG. 83. Comparison of revised model predictions using RG 1.145 (with stability class calculated from either delta T or sigma theta) with measurements for the dose rate (μR/h) from ^{41}Ar at station A117, for 27–28 July 2008 (midnight to midnight).

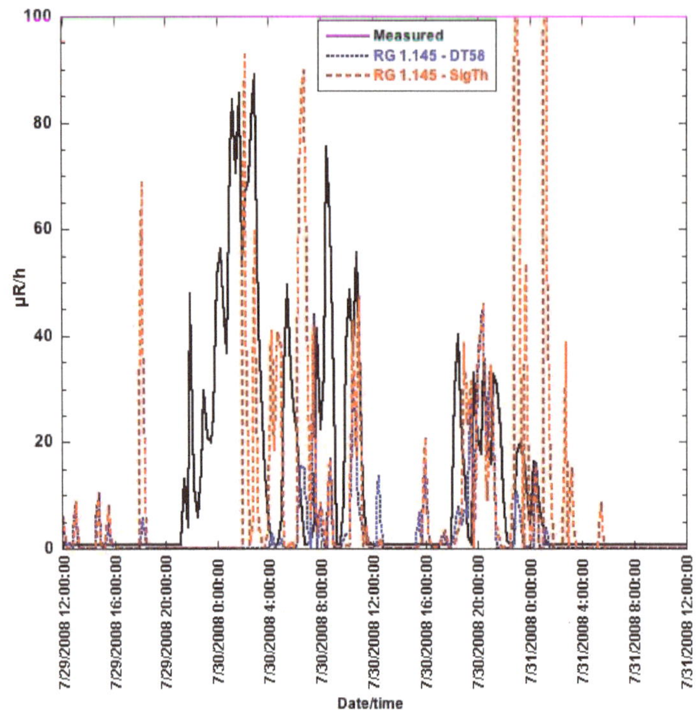

FIG. 84. Comparison of initial model predictions using RG 1.145 (with stability class calculated from either delta T or sigma theta) with measurements for the dose rate (μR/h) from ^{41}Ar at station A117, for 29–31 July 2008 (noon to noon).

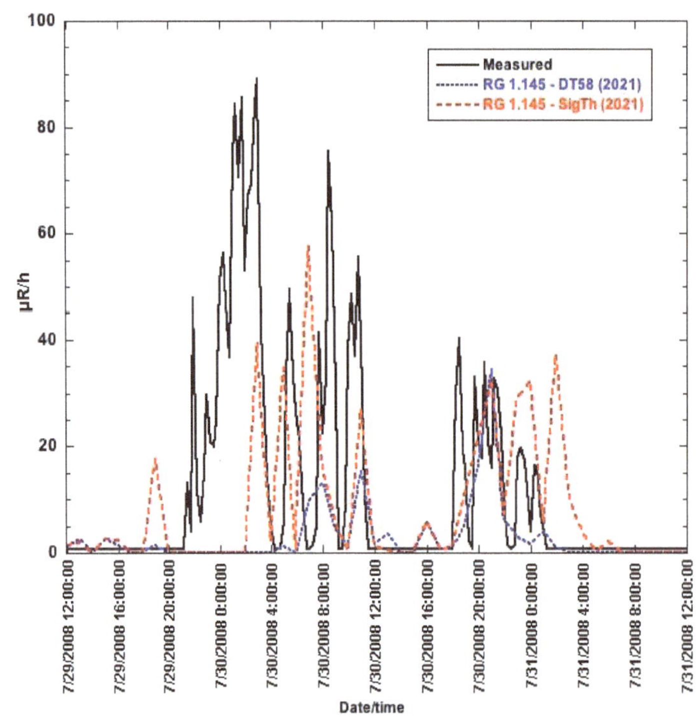

FIG. 85. Comparison of revised model predictions using RG 1.145 (with stability class calculated from either delta T or sigma theta) with measurements for the dose rate (μR/h) from ^{41}Ar at station A117, for 29–31 July 2008 (noon to noon).

6.8.1.4. SPRAY model

Results from four simulations with the SPRAY model for the selected time periods are shown in Figs 86–93. The SPRAY model was run with two different sets of meteorological input information. One set consisted of the meteorological measurements provided to the participants (diagnostic approach); the second set consisted of meteorological forecast data from the Weather Research and Forecasting model (WRF; prognostic approach). In addition, the first set included some WRF data that were essential for the model run but were not available in the set of measurements.

An initial set of model predictions using both approaches was submitted early in the exercise (Figs 86, 88, 90 and 92); these predictions were reported for 30 minute time intervals (corresponding to the meteorological measurements). A second set of model predictions consisted of hourly averages of the 30 minute predictions from the first set of model predictions (Figs 87, 89, 91 and 93). As shown in the figures, the revised predictions generally consisted of shorter, broader peaks, although the difference was not as pronounced as with the RG 1.145 simulations.

The prognostic approach produced peaks during 17–18 July 2008, 20–21 July 2008, 23 July 2008, and 29–31 July 2008 (Figs 65, 86, 87, 92 and 93), that generally did not coincide with the peaks produced by the diagnostic approach (Fig. 64). Two of the peaks from the prognostic approach coincided approximately with measured peaks during 29–31 July 2008 (Figs 92 and 93). These results suggest that the differences in meteorological input information between the measurements and forecasts used in these simulations were significant.

Both the diagnostic and prognostic approach produced peak dose rates during 17–18 July 2008 that were not observed in the measurements (Figs 64, 65, 86 and 87). The diagnostic approach also produced peaks that were not observed in the measurements during other time periods (Figs 88–91) where the prognostic approach did not produce any peaks, while also producing a few peaks that did coincide with measured peaks (e.g. 16:00 and 20:00–00:00 on 27 July 2008; Fig. 91). During 29–31 July 2008, both approaches produced a few peaks not seen in the measurements while also producing peaks that did coincide with some of the peaks in the measurements (Figs 92 and 93).

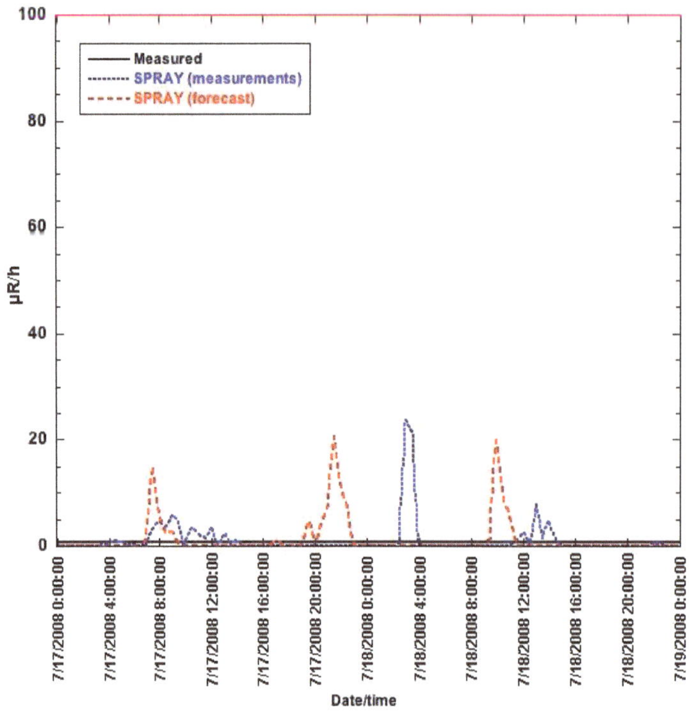

FIG. 86. Comparison of initial model predictions using SPRAY (with either measured or forecast meteorology) with measurements for the dose rate (µR/h) from ^{41}Ar at station A117, for 17–18 July 2008 (midnight to midnight).

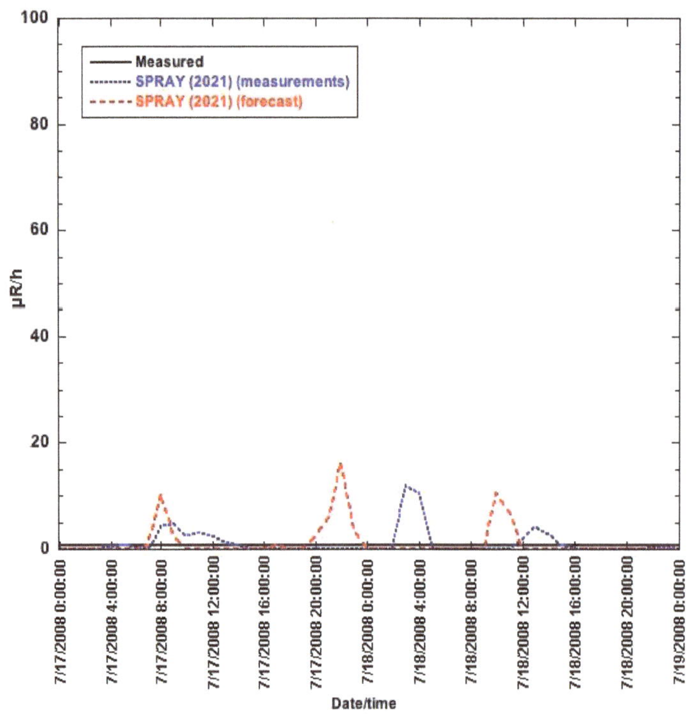

FIG. 87. Comparison of revised model predictions using SPRAY (with either measured or forecast meteorology) with measurements for the dose rate (µR/h) from ^{41}Ar at station A117, for 17–18 July 2008 (midnight to midnight).

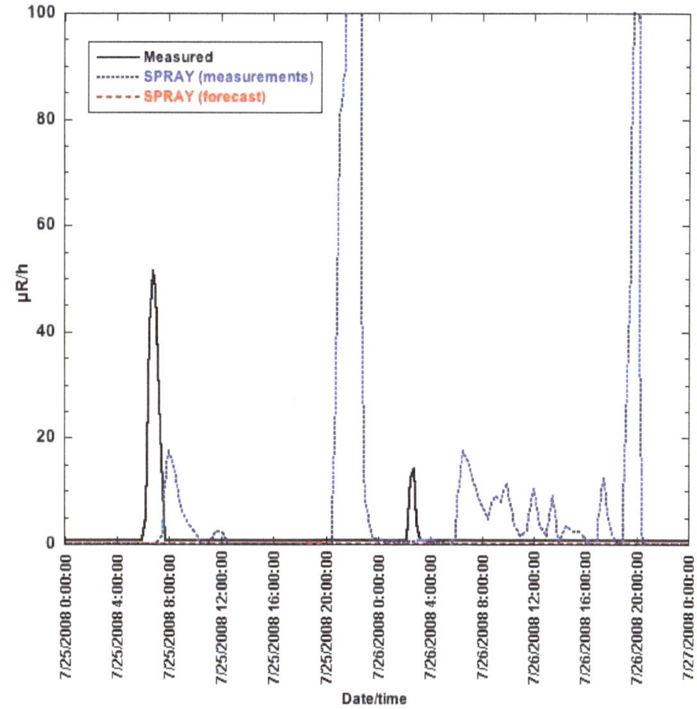

FIG. 88. Comparison of initial model predictions using SPRAY (with either measured or forecast meteorology) with measurements for the dose rate (μR/h) from ^{41}Ar at station A117, for 25–26 July 2008 (midnight to midnight).

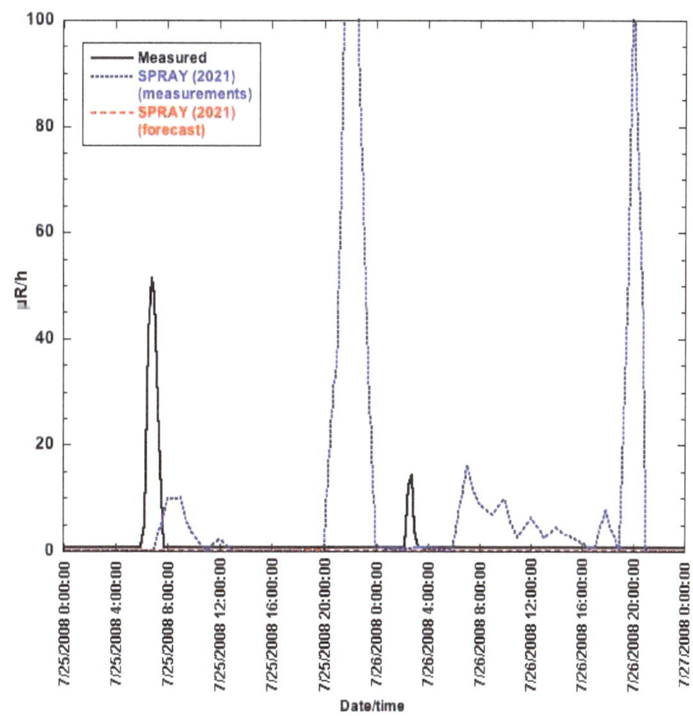

FIG. 89. Comparison of revised model predictions using SPRAY (with either measured or forecast meteorology) with measurements for the dose rate (μR/h) from ^{41}Ar at station A117, for 25–26 July 2008 (midnight to midnight).

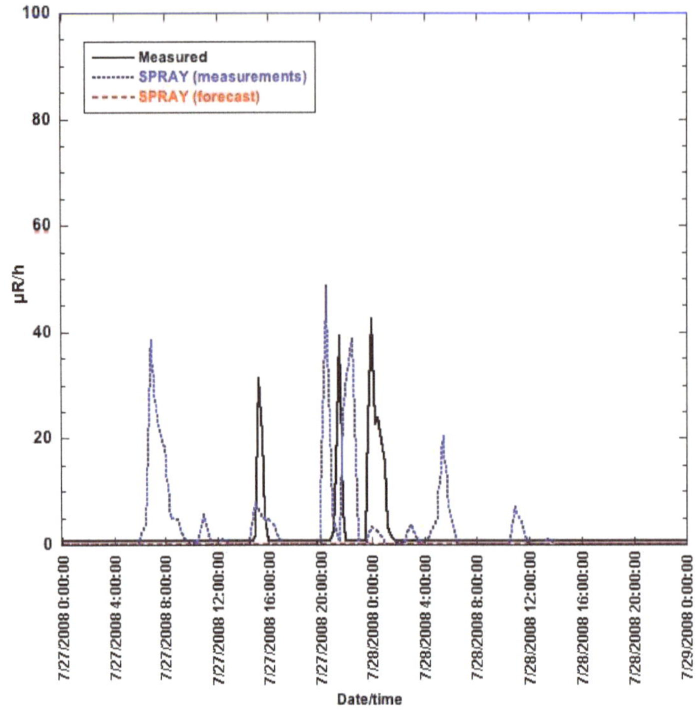

FIG. 90. Comparison of initial model predictions using SPRAY (with either measured or forecast meteorology) with measurements for the dose rate (μR/h) from ^{41}Ar at station A117, for 27–28 July 2008 (midnight to midnight).

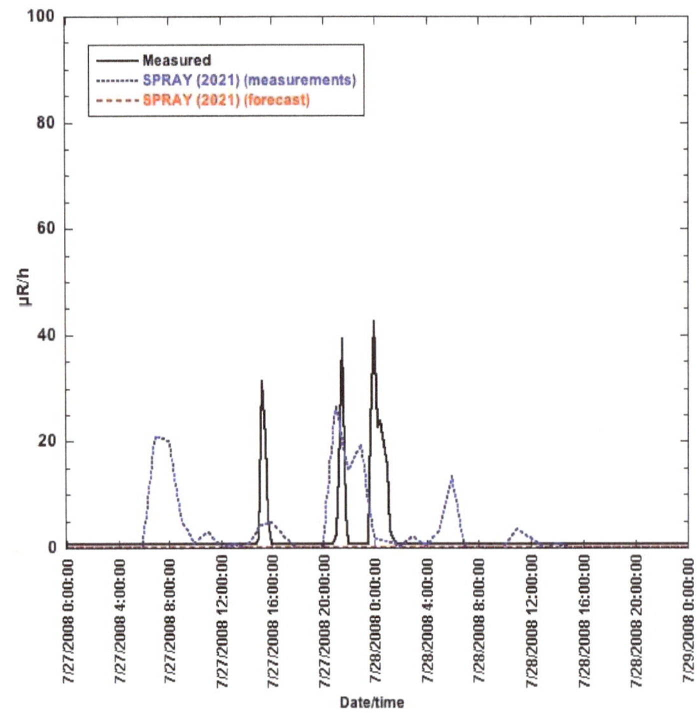

FIG. 91. Comparison of revised model predictions using SPRAY (with either measured or forecast meteorology) with measurements for the dose rate (μR/h) from ^{41}Ar at station A117, for 27–28 July 2008 (midnight to midnight).

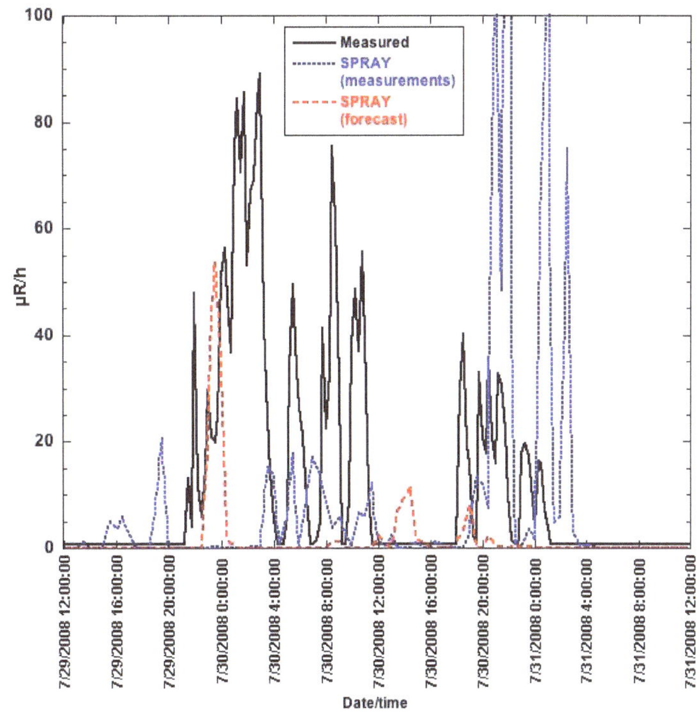

FIG. 92. Comparison of initial model predictions using SPRAY (with either measured or forecast meteorology) with measurements for the dose rate (µR/h) from ^{41}Ar at station A117, for 29–31 July 2008 (noon to noon).

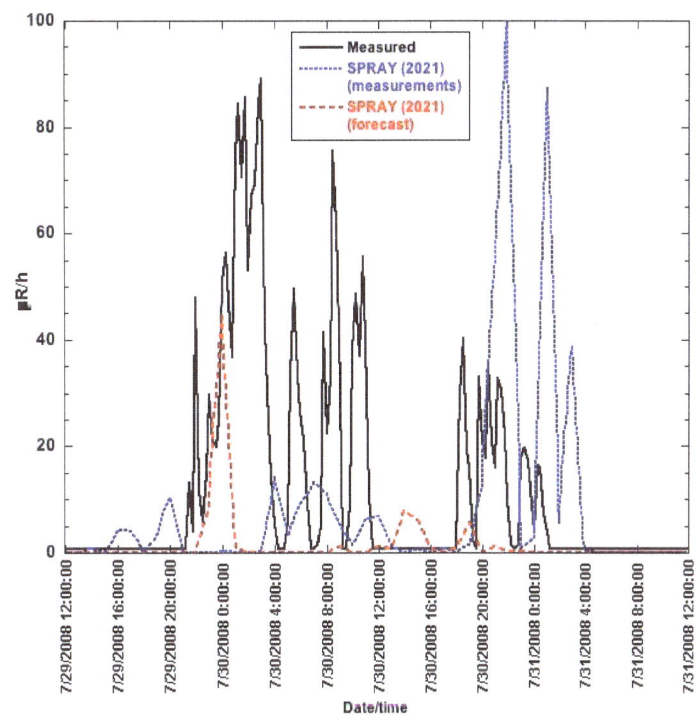

FIG. 93. Comparison of revised model predictions using SPRAY (with either measured or forecast meteorology) with measurements for the dose rate (µR/h) from ^{41}Ar at station A117, for 29–31 July 2008 (noon to noon).

6.8.1.5. *JRODOS decision support system*

Five simulations were performed with the decision support system JRODOS, four using LASAT as the atmospheric dispersion model and one using RIMPUFF for that purpose (Table 12). For both LASAT (Fig. 66) and RIMPUFF (Fig. 67), the simulations used 1 hour averages of the measured meteorological data (1 hour time step) and a 38 km domain. Additional simulations with LASAT used a 15 minute time step (Fig. 68), a 5 km domain, or meteorological forecasts (Meteorological Archival and Retrieval System, MARS, from the European Centre for Medium-Range Weather Forecasts; Fig. 69) (Table 12). Figures 94–97 compare the results of LASAT and RIMPUFF for a 1 hour time step and a 38 km domain for selected time periods. Comparisons with LASAT simulations are also provided for a 1 hour versus 15 minute time step (Figs 98–101), a 38 km versus 5 km domain (Figs 102–105), and meteorological measurements versus meteorological forecasts (Figs 106–109).

LASAT and RIMPUFF gave generally similar predictions (Figs 94–97), with RIMPUFF often (but not always) producing higher peaks than LASAT. Both models predicted peak dose rates during the 17–18 July 2008 time period that were not seen in the measurements (Fig. 94). A few additional peaks were predicted by both models but not seen in the measurements, e.g. 20:00–00:00 on 25 July 2008 (Fig. 95), 19:00–21:00 on 26 July 2008 (Fig. 95), and 06:00–09:00 on 27 July 2008 (Fig. 96). A few peaks in the measurements were largely missed by the model predictions, e.g. just before 08:00 on 25 July 2008 (Fig. 95), near 16:00 on 27 July 2008 (Fig. 96), and some of the peaks on 29–30 July 2008 (Fig. 97).

Use of a 15 minute time step instead of a 1 hour time step with LASAT produced very little difference in the model predictions (Figs 98–101). A notable exception is the much smaller peak with a 15 minute time step for the peak between 00:00 and 04:00 on 28 July 2008 (Fig. 100). Use of a 5 km domain (i.e. domain size: (2 × 4.8 km) × (2 × 4.8 km)) instead of a 38 km domain (i.e. domain size: (2 × 38.4 km) × (2 × 38.4 km)) with LASAT produced essentially no difference in the model predictions (Figs 102–105).

Large differences were observed between the results obtained with LASAT using meteorological measurements (diagnostic approach) and meteorological forecasts (prognostic approach) (Figs 106–109). The results using the MARS forecast data showed two small peaks on 17 July 2008 and nothing during the other time periods (Figs 69, 106–109). These results indicate considerable difference between the MARS forecast data and both the WRF forecast data and the meteorological measurements (Figs 65, 86–93).

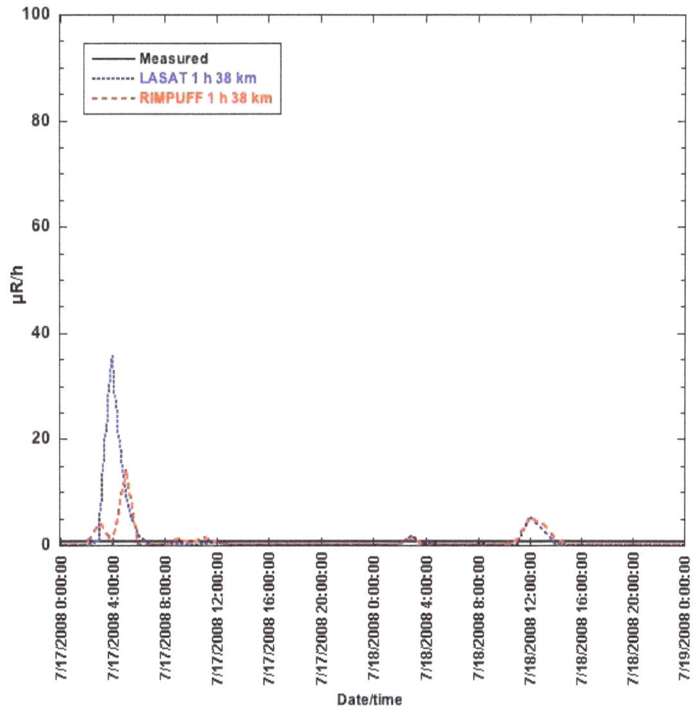

FIG. 94. Comparison of model predictions from JRODOS using either LASAT or RIMPUFF (both with 1 hour time step and 38 km domain) with measurements for the dose rate (µR/h) from ^{41}Ar at station A117, for 17–18 July 2008 (midnight to midnight). Dose conversion: 1 mSv ≈ 0.1 R ≈ 100000 µR.

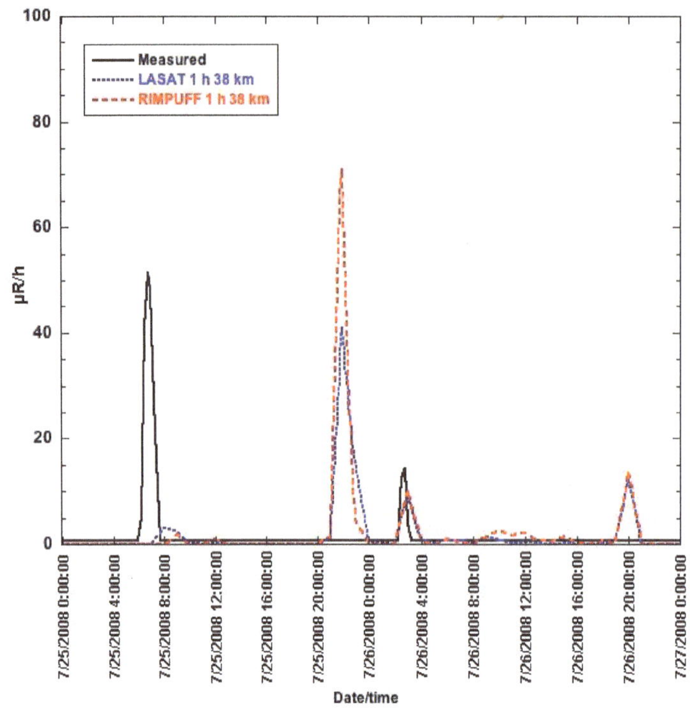

FIG. 95. Comparison of model predictions from JRODOS using either LASAT or RIMPUFF (both with 1 hour time step and 38 km domain) with measurements for the dose rate (µR/h) from ^{41}Ar at station A117, for 25–26 July 2008 (midnight to midnight). Dose conversion: 1 mSv ≈ 0.1 R ≈ 100000 µR.

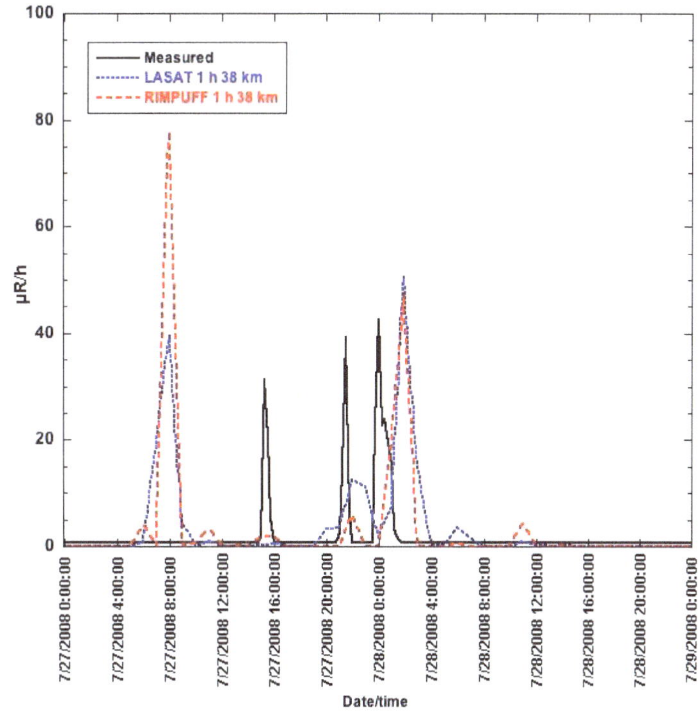

FIG. 96. Comparison of model predictions from JRODOS using either LASAT or RIMPUFF (both with 1 hour time step and 38 km domain) with measurements for the dose rate (µR/h) from ^{41}Ar at station A117, for 27–28 July 2008 (midnight to midnight). Dose conversion: 1 mSv ≈ 0.1 R ≈ 100000 µR.

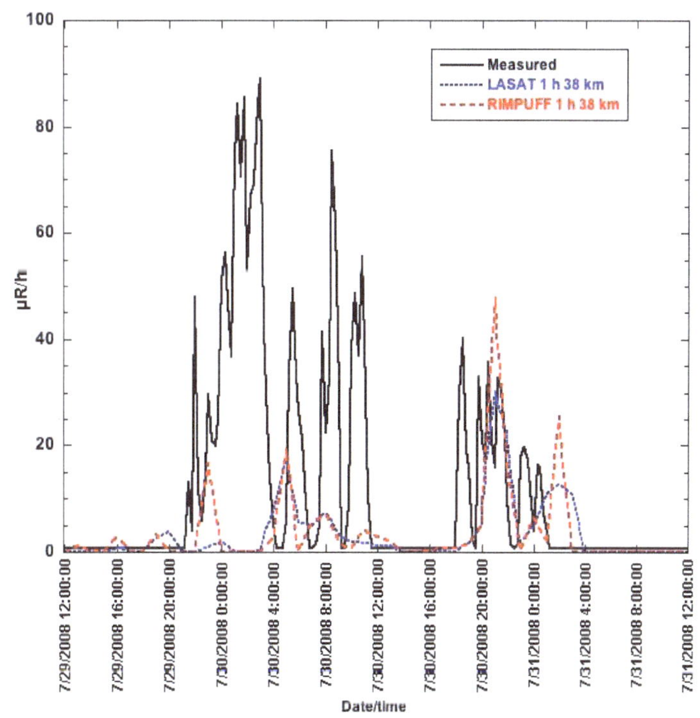

FIG. 97. Comparison of model predictions from JRODOS using either LASAT or RIMPUFF (both with 1 hour time step and 38 km domain) with measurements for the dose rate (µR/h) from ^{41}Ar at station A117, for 29–31 July 2008 (noon to noon). Dose conversion: 1 mSv ≈ 0.1 R ≈ 100000 µR.

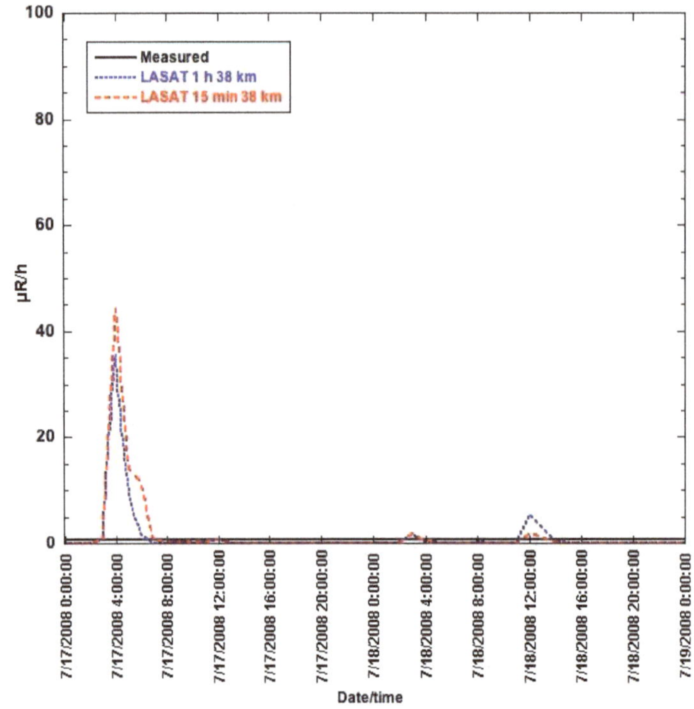

FIG. 98. Comparison of model predictions from JRODOS using LASAT (1 hour or 15 minute time step; 38 km domain) with measurements for the dose rate (μR/h) from ^{41}Ar at station A117, for 17–18 July 2008 (midnight to midnight). Dose conversion: 1 mSv ≈ 0.1 R ≈ 100000 μR.

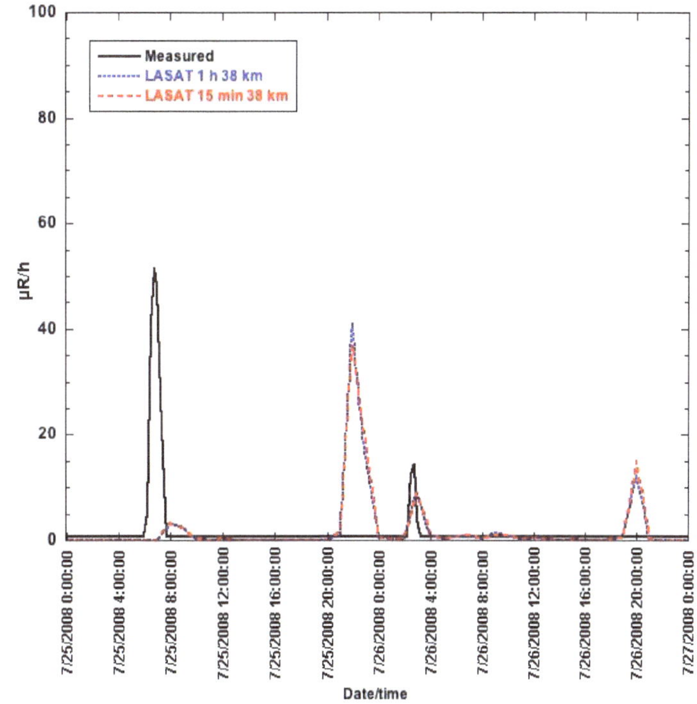

FIG. 99. Comparison of model predictions from JRODOS using LASAT (1 hour or 15 minute time step; 38 km domain) with measurements for the dose rate (µR/h) from ^{41}Ar at station A117, for 25–26 July 2008 (midnight to midnight). Dose conversion: 1 mSv ≈ 0.1 R ≈ 100000 µR.

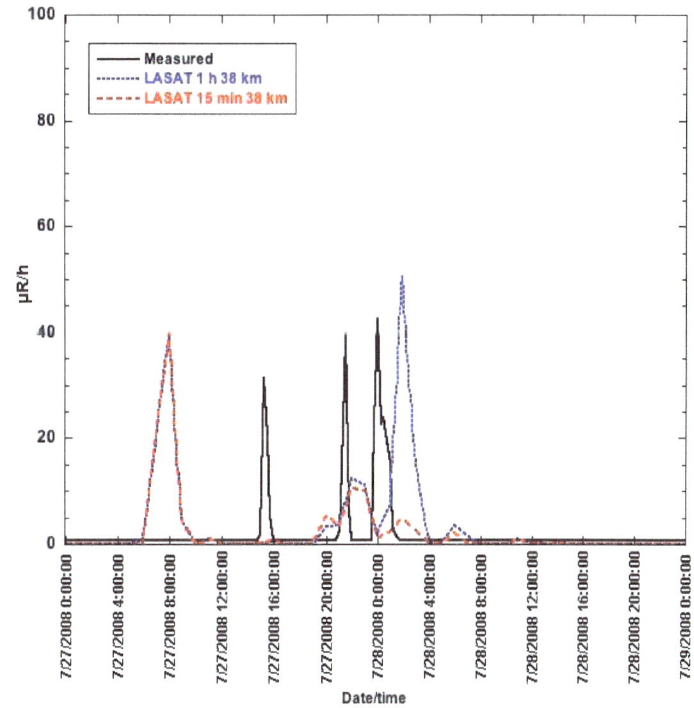

FIG. 100. Comparison of model predictions from JRODOS using LASAT (1 hour or 15 minute time step; 38 km domain) with measurements for the dose rate (µR/h) from ^{41}Ar at station A117, for 27–28 July 2008 (midnight to midnight). Dose conversion: 1 mSv ≈ 0.1 R ≈ 100000 µR.

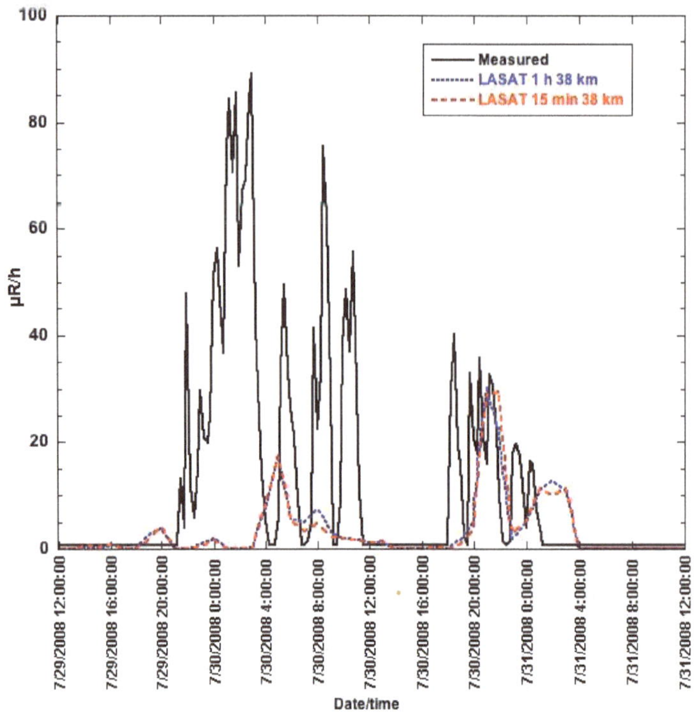

FIG. 101. Comparison of model predictions from JRODOS using LASAT (1 hour or 15 minute time step; 38 km domain) with measurements for the dose rate (μR/h) from ^{41}Ar at station A117, for 29–31 July 2008 (noon to noon). Dose conversion: 1 mSv ≈ 0.1 R ≈ 100000 μR.

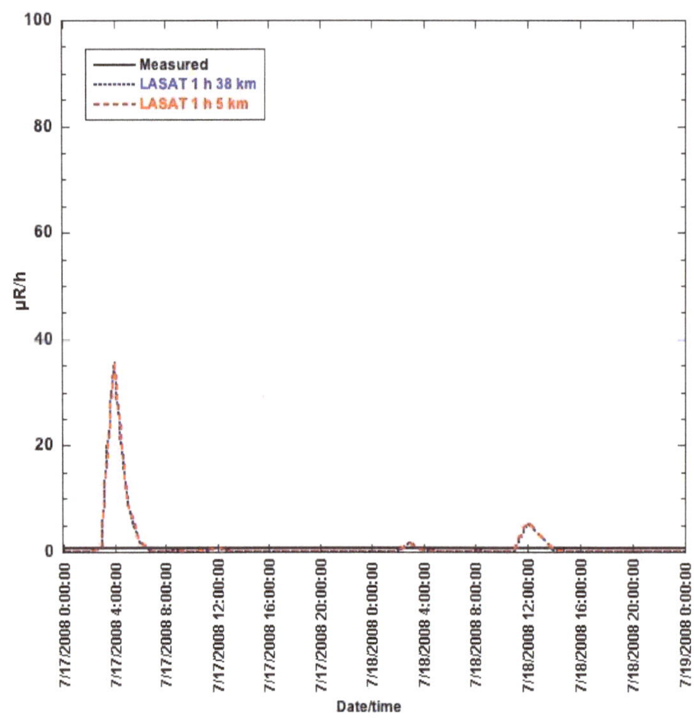

FIG. 102. Comparison of model predictions from JRODOS using LASAT (1 hour time step; 38 km or 5 km domain) with measurements for the dose rate (μR/h) from ^{41}Ar at station A117, for 17–18 July 2008 (midnight to midnight). Dose conversion: 1 mSv ≈ 0.1 R ≈ 100000 μR.

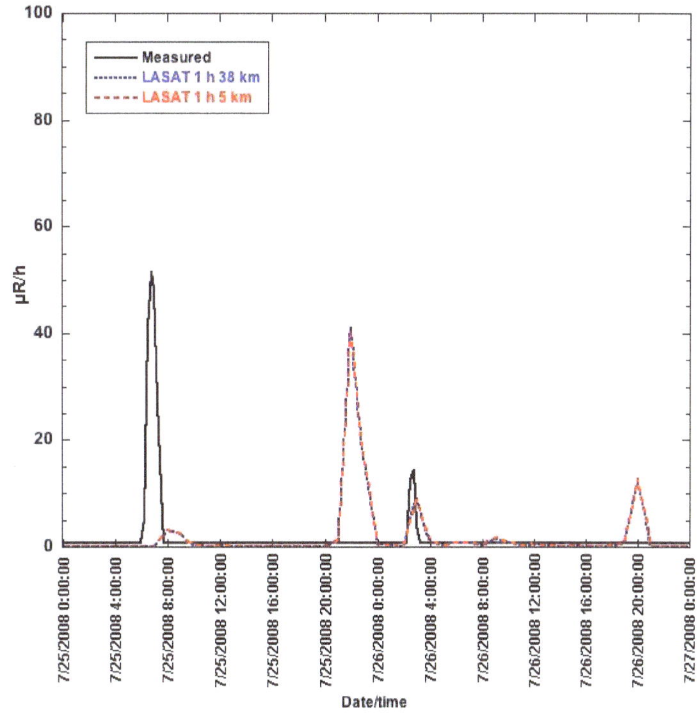

FIG. 103. Comparison of model predictions from JRODOS using LASAT (1 hour time step; 38 km or 5 km domain) with measurements for the dose rate (μR/h) from 41Ar at station A117, for 25–26 July 2008 (midnight to midnight). Dose conversion: 1 mSv ≈ 0.1 R ≈ 100000 μR.

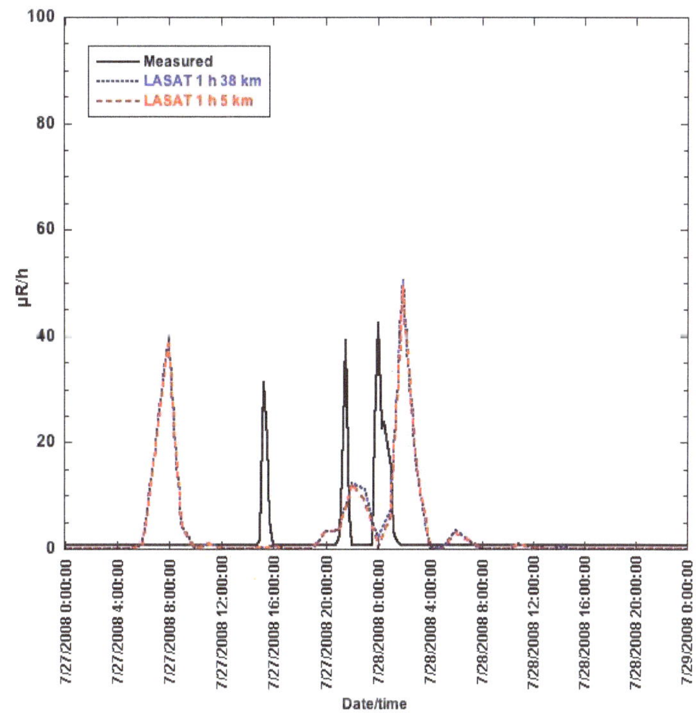

FIG. 104. Comparison of model predictions from JRODOS using LASAT (1 hour time step; 38 km or 5 km domain) with measurements for the dose rate (μR/h) from ^{41}Ar at station A117, for 27–28 July 2008 (midnight to midnight). Dose conversion: 1 mSv ≈ 0.1 R ≈ 100000 μR.

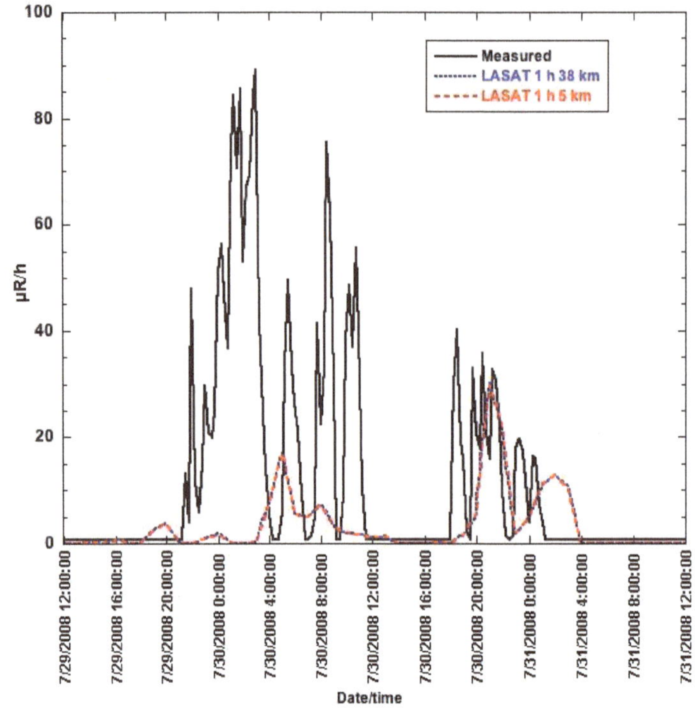

FIG. 105. Comparison of model predictions from JRODOS using LASAT (1 hour time step; 38 km or 5 km domain) with measurements for the dose rate (µR/h) from 41Ar at station A117, 29–31 July 2008 (noon to noon). Dose conversion: 1 mSv ≈ 0.1 R ≈ 100000 µR.

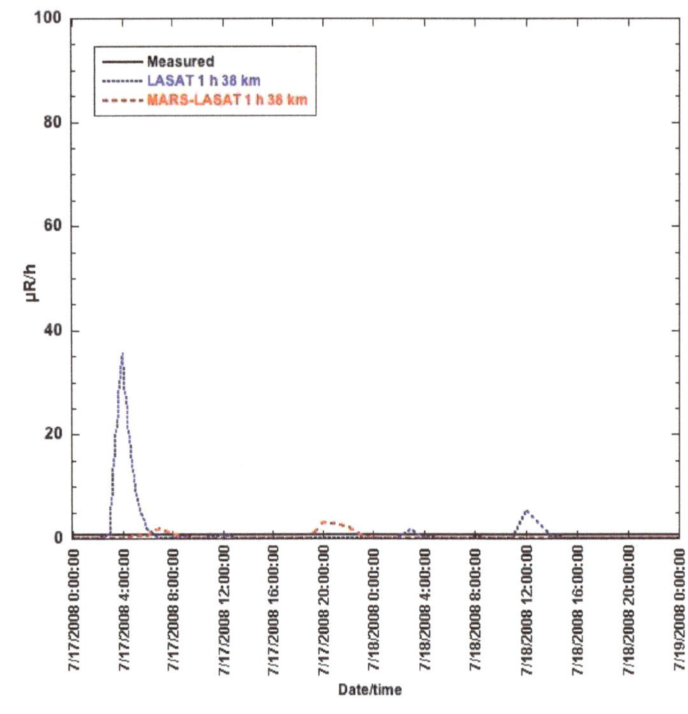

FIG. 106. Comparison of model predictions from JRODOS using LASAT (measured meteorology) or MARS-LASAT (forecast meteorology) (both with 1 hour time step and 38 km domain) with measurements for the dose rate (µR/h) from ^{41}Ar at station A117, for 17–18 July 2008 (midnight to midnight). Dose conversion: 1 mSv ≈ 0.1 R ≈ 100000 µR.

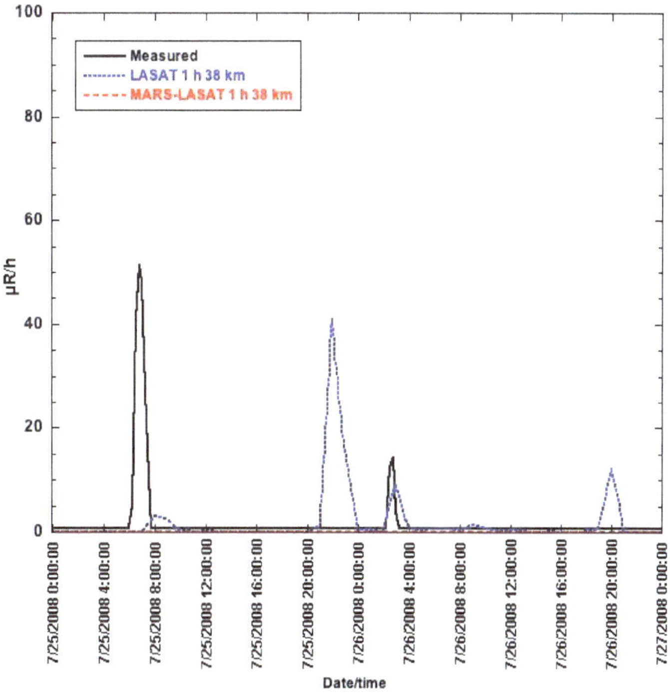

FIG. 107. Comparison of model predictions from JRODOS using LASAT (measured meteorology) or MARS-LASAT (forecast meteorology) (both with 1 hour time step and 38 km domain) with measurements for the dose rate (μR/h) from ^{41}Ar at station A117, for 25–26 July 2008 (midnight to midnight). Dose conversion: 1 mSv ≈ 0.1 R ≈ 100000 μR.

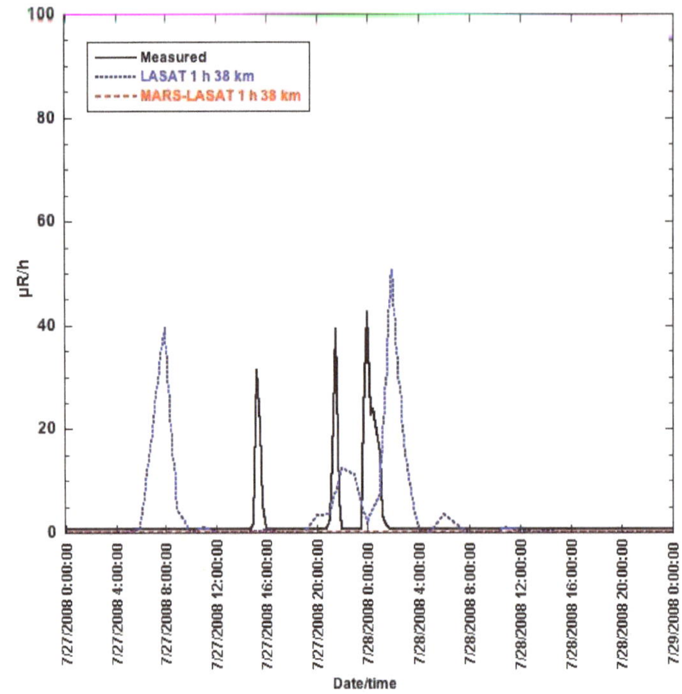

FIG. 108. Comparison of model predictions from JRODOS using LASAT (measured meteorology) or MARS-LASAT (forecast meteorology) (both with 1 hour time step and 38 km domain) with measurements for the dose rate (μR/h) from ^{41}Ar at station A117, for 27–28 July 2008 (midnight to midnight). Dose conversion: 1 mSv ≈ 0.1 R ≈ 100000 μR.

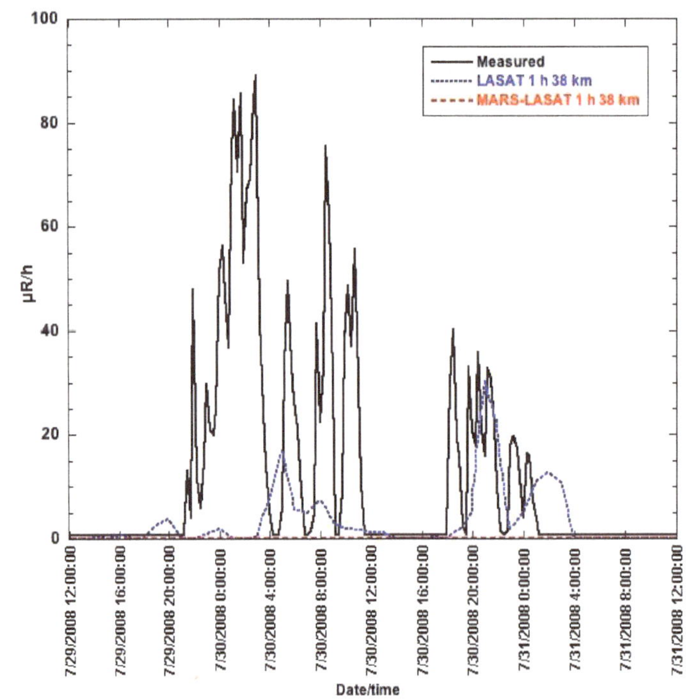

FIG. 109. Comparison of model predictions from JRODOS using LASAT (measured meteorology) or MARS-LASAT (forecast meteorology) (both with 1 hour time step and 38 km domain) with measurements for the dose rate (μR/h) from ^{41}Ar at station A117, for 29–31 July 2008 (noon to noon). Dose conversion: 1 mSv ≈ 0.1 R ≈ 100000 μR.

6.8.2. Time series of ^{41}Ar activity concentrations in air (Bq/m^3)

The requested endpoints for the modelling exercise did not specifically include a time series of ^{41}Ar activity concentrations in air. Some, if not most models, calculated the activity concentration in air and then converted it to a dose rate. For example, with both RG 1.145 and SPRAY, the predicted activity concentrations of ^{41}Ar in air were converted to dose rate using the factor 2.20×10^{-10} Sv/h per Bq/m^3 (Table 11).

The predictions submitted using JRODOS included both the activity concentrations in air and the dose rates (Fig. 110). In this case, the dose rate appears to have been calculated a little differently than a simple multiple of the activity concentration in air; when plotted together, the two plots are similar but not quite identical. Most likely, in its dose calculation, JRODOS includes the activity concentration in the 'cell' at the receptor location plus the contribution to dose from the activity concentration in adjacent cells.

6.8.3. Time integrated endpoints

As mentioned earlier, comparison of time dependent endpoints is subject to uncertainty in timing of model predictions for the endpoint. One useful method of comparing model predictions and measurements of time dependent endpoints for a given location is by comparison of time integrated endpoints, which can reduce the effects of small differences in the predicted time dependence. For the Chalk River scenario, time integration of predicted or measured dose rates (dose rate × time) results in an estimate of dose to a hypothetical receptor at the location of the monitoring station (Fig. 111).

Several observations can be made from this comparison. Both simulations based on meteorological forecasts gave substantially lower time integrated results than their corresponding simulations based on measured meteorology (SPRAY forecast versus Spray measured; MARS-LASAT versus LASAT). The three simulations with LASAT and measured meteorology gave very similar results to each other and to the results using RIMPUFF. Similar results were obtained using the sigma theta version of RG 1.145 and using SPRAY with measured meteorology; these results were somewhat higher (by a factor of 1.24–1.35) than the time integration of the measured dose rates, while the other models gave results lower than the time integration of the measured dose rates, in most cases by a factor of 0.25–0.6. The result using MARS-LASAT (forecast meteorological data) was about 1% of the result based on measured dose rates, consistent with the very few peaks predicted in this simulation (Section 6.8.1.5).

6.8.4. Contour maps of ^{41}Ar dose rates

Contour maps of predicted dose rates from JRODOS (LASAT, 1 h time step, 38 km or 5 km domain), RG 1.145, and SPRAY and are shown in Figs 112 and 113 for several selected time points. Use of contour maps addresses some of the uncertainty in location for the predictions described in Section 6.8.1 (see also Section 4.4.1).

For JRODOS (LASAT) there was very little difference in the time series for the specific receptor location between the 38 km and 5 km domains (Figs 102–105). The contour maps show similar dose rates at the receptor location (Fig. 112), but the overall extent of the plumes varies with the domain size.

With SPRAY, the differences in results between the diagnostic and prognostic approaches are evident, as discussed elsewhere. For RG 1.145, the differences in results between the two approaches for determining the stability class, while evident, are generally not great. The exception is the results for 25 July 2008 07:00, for which a plume is visible with the sigma–theta approach but not with the delta T 58 approach.

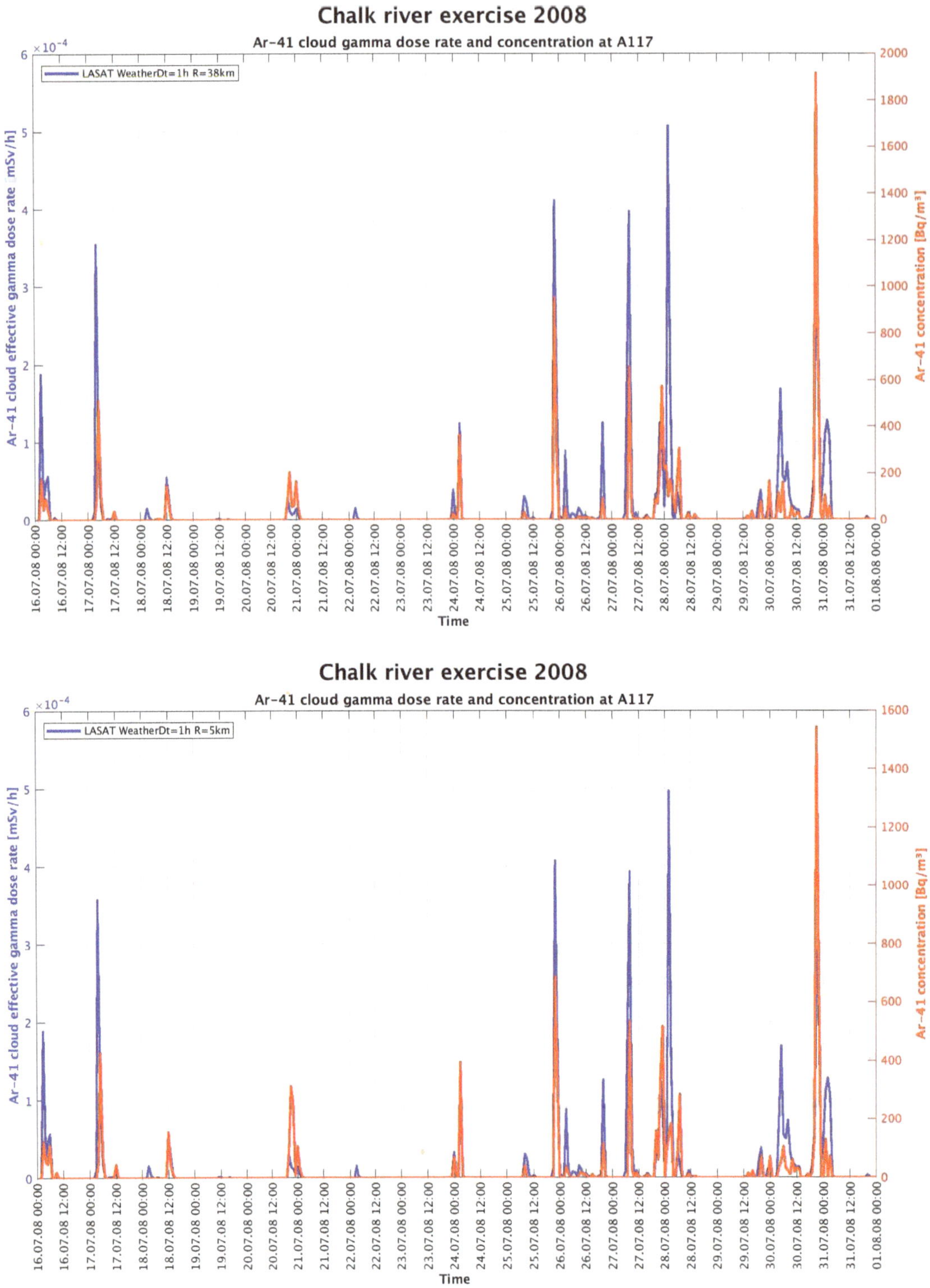

FIG. 110. Model predictions from JRODOS using LASAT (measured meteorology, 1 hour time step; 38 km domain (top) or 5 km domain (bottom)) comparing the predicted activity concentration of ^{41}Ar in air (red, y-axis on right, Bq/m^3) and the predicted gamma dose rate (blue, y-axis on left, mSv/h). Dose conversion: 1 mSv ≈ 0.1 R ≈ 100000 μR.

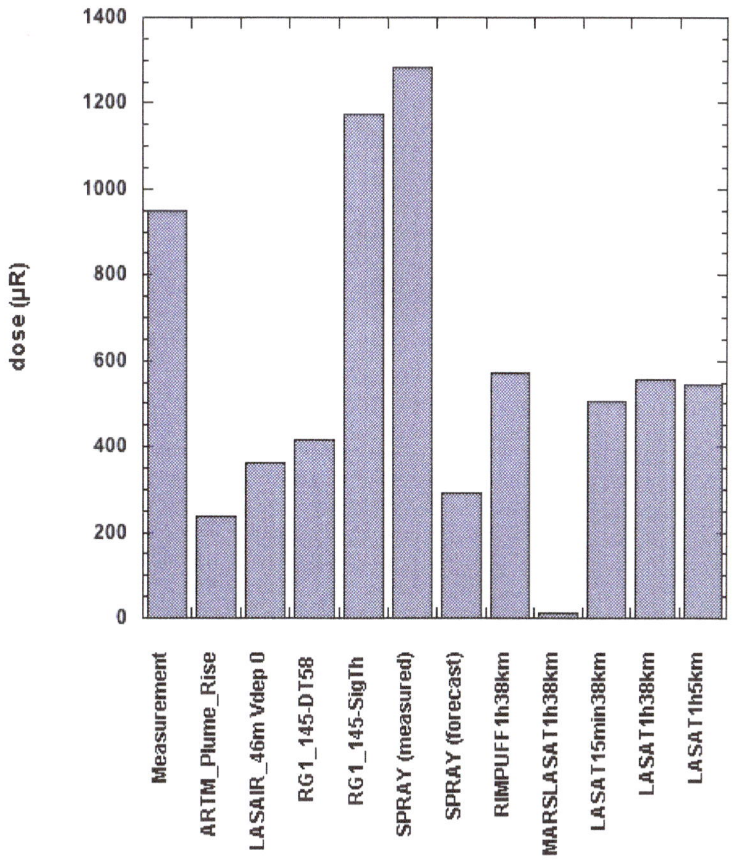

FIG. 111. Comparison of dose estimates for the Chalk River exercise, based on time integration of measured and predicted dose rates Dose conversion: 1 mSv ≈ 0.1 R ≈ 100000 μR.

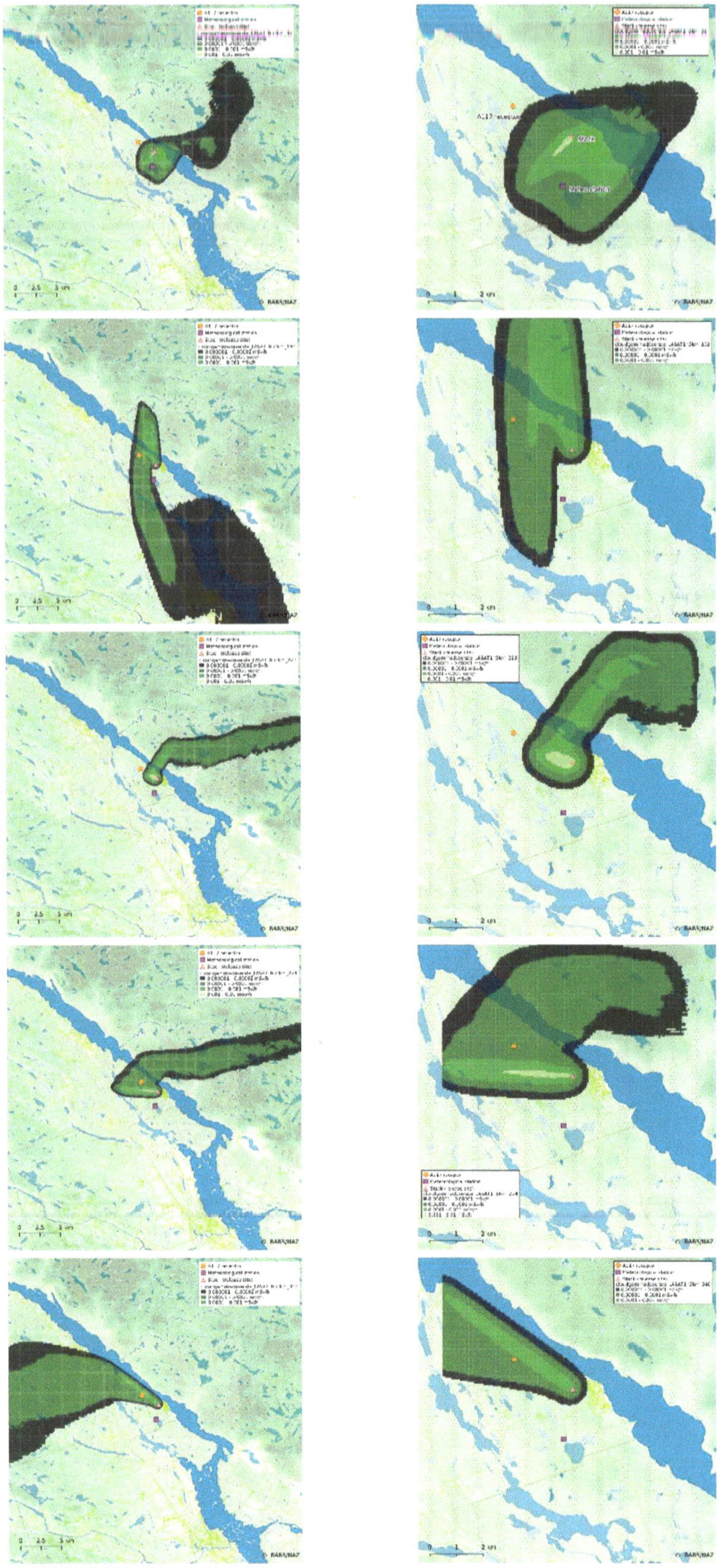

FIG. 112. Contour maps of predicted dose rates(mSv/h) from LASAT (1 h time step), with 38 km domain (left) or 5 km domain (right). From top to bottom, predictions are shown for 32 h, 192 h, 223 h, 224 h, and 340 h after the start (corresponding times EST: 17 July 2008 08:00, 24 July 2008 00:00, 25 July 2008 07:00, 25 July 2008 08:00, 30 July 2008 04:00). Note that the scales are different between the left and right columns. Dose conversion: 1 mSv ≈ 0.1 R ≈ 100000 µR.

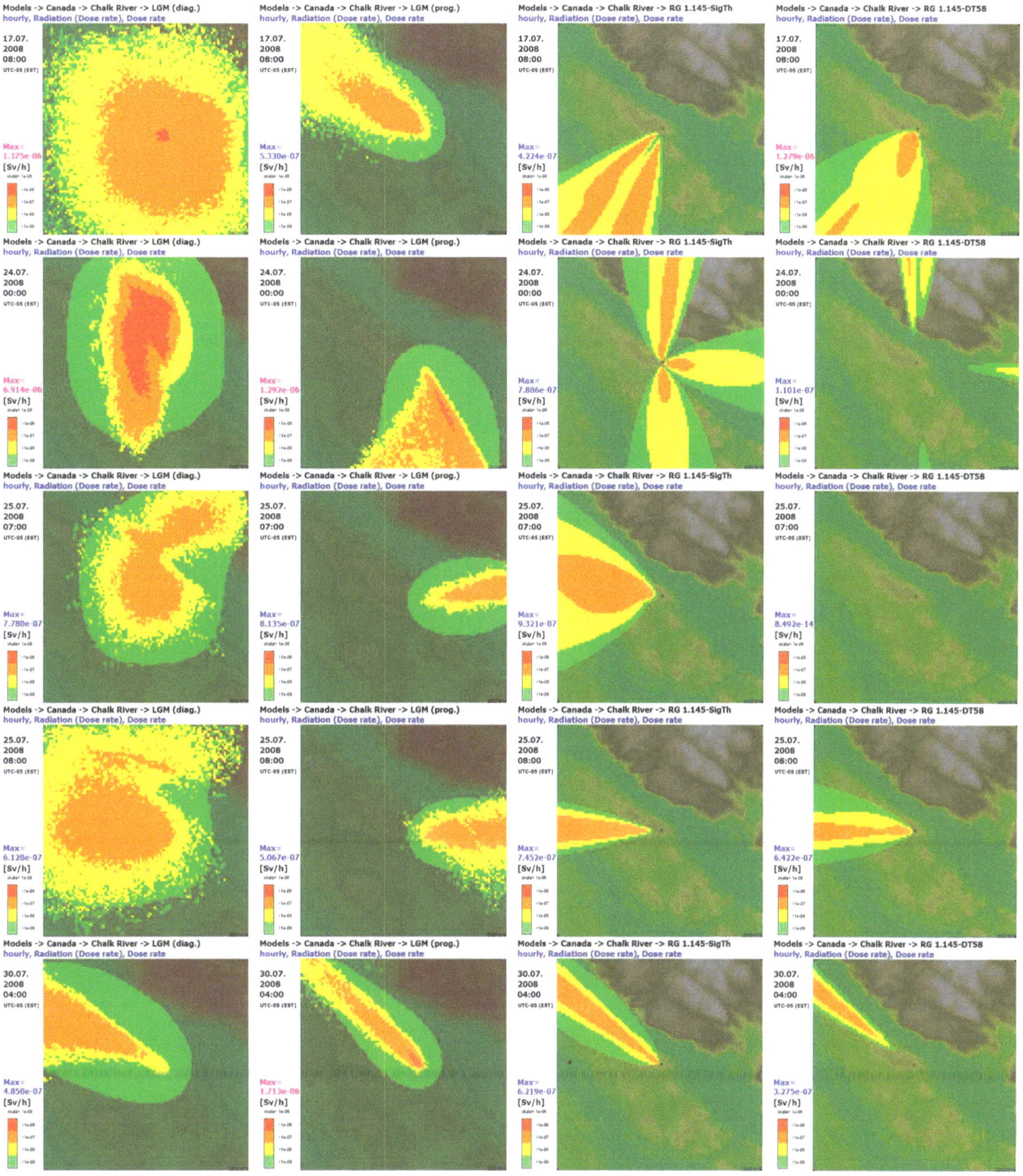

FIG. 113. Contour maps of predicted dose rates (Sv/h) from SPRAY and RG 1.145. From left, SPRAY in diagnostic mode, SPRAY in prognostic mode, RG 1.145 with sigma–theta approach, and RG 1.145 with Delta T 58 approach. From top to bottom, predictions are shown for 17 July 2008 08:00, 24 July 2008 00:00, 25 July 2008 07:00, 25 July 2008 08:00, and 30 July 2008 04:00 (all times EST). Dose conversion: 1 Sv ≈ 100 R ≈ 10^8 μR.

6.9. FINDINGS FROM THE CHALK RIVER EXERCISE

Perhaps most noteworthy for general discussion is the large difference between simulations of time dependent dose rates with a given model using measured meteorology versus forecast meteorology, as described above for SPRAY and LASAT. This was seen with forecast meteorology from two sources, giving somewhat different results. In general, model predictions of dose rate using forecast data showed many fewer peaks than the measurements of dose rate. Predictions of dose rate using measured meteorology tended to show more peaks than did the measurements of dose rate.

Also, as described above in regard to individual models, the various sets of predictions included peaks in the predicted dose rate at times when there were no peaks in the measurements ('false alarms'). For example, during the time period 17–18 July 2008, no peaks are visible in the measurements, but every model simulation (including the simulations with forecast rather than measured meteorology) included peaks during that period, although not in the same places or of the same magnitude. Another example is 27 July 2008, for which all simulations produced a peak not seen in the measurements, with the exception of the JRODOS simulation using forecast meteorology. In contrast, the measurements for 29–31 July 2008 included more peaks and larger peaks than did the predictions with any model.

One possible explanation for these observations is the distance between the meteorological station and either the emissions source or the monitoring (receptor) location. Ideally, the meteorological measurements need to be obtained as close as possible to the release location. As described in Section 4 of this publication and in Ref. [11], prediction of an activity concentration or other atmospheric dispersion endpoint at a point location and specified time involves uncertainty in both location and time, especially if the meteorological data are not obtained at ideal locations with respect to the source and the receptor location. As described in Sections 6.8.3 and 6.8.4, comparisons of time integrated results can be useful with respect to uncertainty in timing, and comparisons of contour plots can be useful with respect to uncertainty in location.

Another consideration is the availability of the ^{41}Ar dataset, which was provided as "reported releases of ^{41}Ar during the time period of 16 July 2008 to 6 August 2008 between 2×10^{14} and 3.5×10^{14} Bq/week" (see Section 6.3). As no other information was available, it was up to the participants to define the correlation of the weekly release number to the corresponding time interval. For example, the release (value) could be defined at the beginning, in the middle or at the end of the time interval, which causes different results for the predicted concentration downwind.

A related possible cause of the discrepancies is the uncertainty of the source term, which was provided only on a weekly basis. It is not known whether any fluctuation of the emission flux occurred within a week or if it remained constant. It is conceivable that on 17 July 2008 no emission took place, and between 29 and 31 July 2008 the emitted ^{41}Ar flux was higher than the weekly average.

In the simulations with JRODOS, it is apparent that the differences in predictions between the two atmospheric dispersion models, LASAT and RIMPUFF, are not large. In addition, with LASAT, there is little difference in the time dependent predictions using two different time steps (15 minutes versus 1 hour), and essentially no difference using a 5 km domain versus a 38 km domain.

As described above, the difference in results between simulations using measured meteorology versus forecast meteorology was large, in the two cases in which the same model was used with two sets of meteorology. In addition, there were relatively consistent differences across models between measured and predicted dose rates at the receptor location. These observations reinforce the importance of having good and sufficient meteorological data for a particular site.

Predicted time series for ^{41}Ar dose rates showed consistencies with each other or with the measurements for some time periods, and some substantial differences for other time periods. When compared in terms of a time integrated endpoint, most simulations gave results within a factor of four or less of the observations (Section 6.8.3). The main exception was the simulation with LASAT using forecast meteorology, which was substantially lower than the LASAT predictions using measured meteorology.

The ARTM, SPRAY, and JRODOS models took terrain effects into account. Given other differences among the models, it is not possible to know whether this explains the differences in model results, although it is probably part of the explanation. Future exercises could consider running some models with and without terrain effects.

7. DOSE ASSESSMENT EXERCISE FOR A CONTAMINATED URBAN AREA

7.1. INTRODUCTION TO THE FUKUSHIMA EXERCISE

The Fukushima modelling exercise was based on measurements made in Japan following the accident at the Fukushima Daiichi Nuclear Power Plant in March 2011 and on surveys carried out by the Japan Atomic Energy Agency (JAEA) [33, 34]. The first stage of the exercise was carried out during the MODARIA I programme [11]. This involved prediction of external dose rates at specified locations in the absence of remediation, and prediction of the external doses to specified reference individuals in the absence of remediation [11]. The second stage of the exercise, conducted during the MODARIA II programme, is described in this section. This stage involved prediction of external radiation doses received by populations in an urban situation, in the absence of remediation. A summary of the results of the exercise are published in Ref. [14]. Future stages of the exercise could involve prediction of the effectiveness of various remediation strategies, including the prediction of external dose rates and doses following specified remediation actions.

The objectives of the Fukushima exercise during the MODARIA II programme were: (1) to compare measured and predicted distributions of external doses to an urban population; and (2) to compare probabilistic and deterministic approaches to assessment of external doses to the representative person[6] .

7.2. DESCRIPTION OF THE EXERCISE

Input information for the Fukushima exercise included environmental information (radionuclide composition, deposition levels, conditions of the initial deposition event) and habit information (information about typical Japanese houses and typical locations in the target area, time spent at various locations). The input information is summarized below. Detailed data on meteorology, dose rates, and deposition levels of ^{137}Cs were provided to participants in electronic form.

7.2.1. General description of Fukushima City

This exercise was set in Fukushima City, which is the capital of Fukushima Prefecture. The prefecture has a population of about 280 000 people in an area of about 760 km^2. Most of the inhabitants reside in an urban area paved with either concrete or asphalt. About 35% of the area is habitable, and about 80% of the inhabitants live in one or two story wooden houses.

Fukushima City is located approximately 60 km northwest of the Fukushima Daiichi Nuclear Power Plant. Following the accident at the Fukushima Daiichi Nuclear Power Plant, concentrations of radioactivity in Fukushima City were not high enough to warrant evacuation of the population, and people continued to live there.

7.2.2. Environmental data

Environmental measurements in Fukushima City and Fukushima Prefecture include background gamma dose rates in the prefecture (i.e. before the accident), gamma dose rates and precipitation measured in Fukushima City soon after the accident, and measured levels of radioactivity on the ground surface and on the roofs and walls of houses after the accident.

---

[6] The representative person is defined as "An individual receiving a dose that is representative of the doses to the more highly exposed individuals in the population" [35].

7.2.2.1. *Background levels of gamma dose rate*

Before the accident in March 2011, the gamma dose rate measured in Fukushima Prefecture was about 0.04 μSv/h [36, 37].

7.2.2.2. *Information about the accident*

Figure 114 shows the relationship between the gamma dose rate [38] and precipitation [39] in Fukushima City following the accident on 11 March 2011. A steep increase in dose rate occurred around 18:00 on 15 March 2011, coinciding with an increase in precipitation. It is assumed that the wet conditions contributed to the radioactive fallout and contamination in Fukushima City. Table 13 summarizes the composition of the deposited radioactivity, relative to that of ^{137}Cs [40].

7.2.2.3. *Contamination of the ground surface*

This exercise made use of the results of airborne monitoring surveys performed in 2012 [38, 39]. Table 14 summarizes the measured contamination level of ^{137}Cs in Fukushima City, decay corrected to 31 May 2012 [14]. Due to the residential and building areas of Fukushima City being located in a basin, and due to the passage of the radioactive plume coinciding with rainfall, higher contamination occurred in the residential and building areas. Thus, the average contamination level of ^{137}Cs in the residential and building areas of Fukushima City was higher than that of ^{137}Cs for all types of land use combined, by about 50% (Table 14).

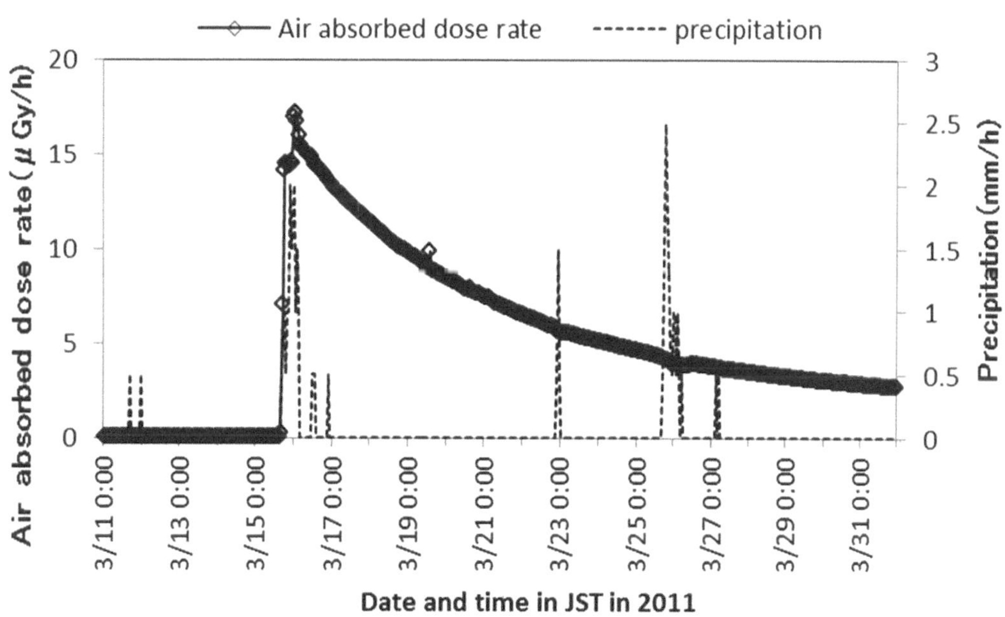

FIG. 114. Relationship between gamma dose rate in air (μGy/h) and precipitation in Fukushima City in March 2011(data from Refs [38, 39].

TABLE 13. COMPOSITION OF DEPOSITED RADIOACTIVITY IN FUKUSHIMA CITY [40][a]

Radionuclides	Deposited activity[b], normalized to ^{137}Cs
^{110m}Ag	0.0028
^{129m}Te	1.1
$^{132}Te + ^{132}I$	8
^{131}I	11.5
^{134}Cs	1
^{136}Cs	0.17
$^{137}Cs + ^{137m}Ba$	1

[a] Table from Ref. [14].
[b] The composition given in this table is decay corrected to 15 March 2011.

TABLE 14. SURFACE DEPOSITION OF ^{137}CS IN FUKUSHIMA CITY[a]

Land use	Correction date	Sample size	Surface density of ^{137}Cs on the ground (kBq m^{-2})					
			AM	SD	GM	GSD	Min	Max
All types of land use	31 May 2012	12 001	100	80	70	2.5	5	460
Residential and building areas	31 May 2012	1103	152	49	144	1.4	35	380

[a] Table from Ref. [14].
AM = arithmetic mean; SD = standard deviation; GM = geometric mean; GSD = geometric standard deviation; Min = minimum value; Max = maximum value.

7.2.2.4. Contamination of roofs and walls of houses

Measurements of radioactivity on the roofs and walls of houses contaminated as a result of the accident at the Fukushima Daiichi nuclear power plant were performed in 2015 in Okuma Town and Tomioka Town [41]. The surface contamination level of ^{137}Cs was measured on 212 roofs and 478 walls. The average relative deposition of ^{137}Cs on the roofs and walls in urban areas were 0.02 ± 0.02 and 0.01 ± 0.005[7], respectively and represent the ratio of the surface contamination level on roofs and walls (i.e. on the date of measurement) to the contamination level on permeable plane fields adjacent to the target building (e.g. backyard, unpaved grass field, or bare ground) on 15 March 2011.

7.2.3. Habit data

The targets of the assessment were: (1) an indoor worker; and (2) and outdoor worker. Both were assumed to live and spend all of their time (work and leisure) in Fukushima City.

7.2.3.1. Behavioural patterns

For both indoor and outdoor workers, the time spent at various places (e.g. inside and outside of the house, workplace, or other places) was obtained from a survey (see Refs [33, 34] and also summarized in Table 15). The survey was performed for the period February 2012 to January 2013 in Fukushima City. The statistics in Table 15 were obtained from the survey data, accounting for seasonal variability. Surveys of behavioural patterns were performed for the same individuals for whom there are actual measurements of individual external doses [14].

[7] The standard deviation of the relative ^{137}Cs deposition on walls was given as "<0.01" in Ref. [41].

7.2.3.2. *Typical location and house*

Information about typical locations and houses for use in modelling was obtained by surveying the locations of persons participating in individual dose measurement and those in the behavioural pattern surveys. Information on the typical location and house structure and material is given in Table 16. Figure 115 shows a schematic drawing of the typical location and model of a Japanese house [42].

Many studies have been performed to evaluate dose reduction effects (shielding factors) for Japanese houses, based on either calculations or actual measurements (summarized in Ref. [14]). The result of a literature review on the dose reduction factor for Japanese wooden houses reported that the factors are mostly around 0.4 [43].

7.2.4. Modelling endpoints for the exercise

Doses were estimated, considering only the contribution from external exposure due to deposited radionuclides. Based on the information provided, for each population group, the following endpoints were assessed:

(1) The distribution of annual effective dose from external exposure due to deposited radionuclides during the period from February 2012 to January 2013 for indoor workers and outdoor workers.
(2) The dose to the representative person for the two population groups, using both a deterministic and a probabilistic approach.

TABLE 15. SURVEY RESULTS FOR TIME SPENT IN VARIOUS PLACES[a,b]

Population group (recommended distribution for time spent outdoors from occupancy survey)		Total (h)		Time spent per day (h)					
				Home		Workplace		Other	
		Indoor	Outdoor	Indoor	Outdoor	Indoor	Outdoor	Indoor	Outdoor
Indoor worker (Lognormal) N = 11	AM	23.3	0.7	16.1	0.3	6.1	0.0	1.1	0.4
	SD	0.47	0.48	1.45	0.38	0.64	0.06	0.9	0.39
	GM	23.3	0.5	16.1	0.3	6.1	0.1	0.9	0.3
	GSD	1.0	2.6	1.1	1.0	1.1	1.0	2.6	2.5
	95%tile	23.9	1.3	18.0	0.9	6.9	0.1	2.4	1.0
Outdoor worker (Normal) N = 33	AM	18.3	5.7	16.6	3.5	0.4	1.4	1.3	0.8
	SD	2.1	2.1	2.5	2.6	0.6	2.2	0.7	1.0
	GM	15.9	7.4	13.9	1.3	1.3	4.4	0.6	0.5
	GSD	1.2	1.4	1.2	3.4	3.4	1.7	2.7	2.6
	95%tile	21.0	8.3	20.1	7.2	1.3	5.0	2.4	2.5

AM = arithmetic mean; SD = standard deviation; GM = geometric mean; GSD = geometric standard deviation; 95%tile = 95th percentile.

[a] Table reproduced from Ref. [14] with permission courtesy of Journal of Radiological Protection.

[b] Occupancy survey data were obtained for the period February 2012 to January 2013 in Fukushima City.

TABLE 16. TYPICAL LOCATION, HOUSE STRUCTURE, AND MATERIAL OF JAPANESE HOUSES[a]

Item	Characteristic	Value
House location	Distance between houses	5 m
	Width of road in front of house	10 m
	Building to land ratio	> 40%
House structure	Number of stories	2
	Construction area	130 m^2
	Eave height from ground	6 m
	Height to the peak of the roof (ridge height)	7.2 m
House material (Mass thickness)	Roof	1.1 g cm^{-2}
	Wall	2.4 g cm^{-2}
	Window	0.75 g cm^{-2}

[a] Table reproduced from Ref. [14] with permission courtesy of Journal of Radiological Protection.

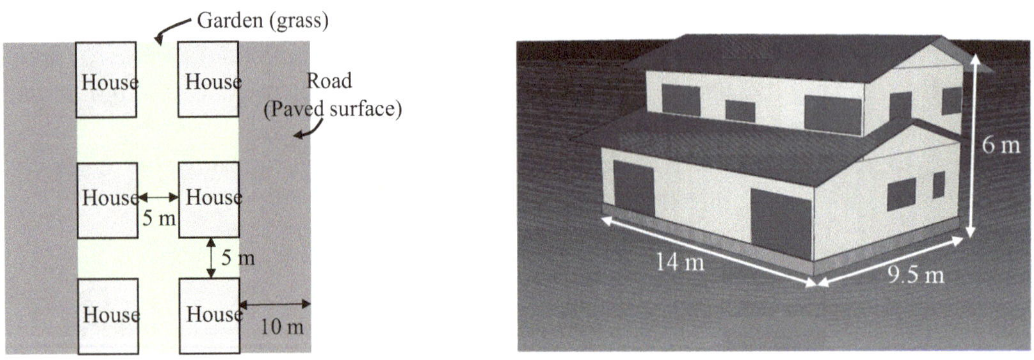

FIG. 115. Schematic drawing of typical location in Fukushima City and the typical model of a Japanese house (figure reproduced from Ref. [14] with permission courtesy of Journal of Radiological Protection).

7.3. MODELS USED IN THE EXERCISE

Five individuals or groups participated in the modelling exercise, using five different models, as summarized in Table 17 [14]. Additional information on the application of TINT-UNSCEAR, ERMIN and METRO-K to this exercise are given in Appendix I.

Two major types of modelling approach were used [14]. Two models (DPRO and TINT-UNSCEAR) used a location based approach, which is based on location specific functions related to changes in dose rates to estimate the dose rates in the locations of interest. The other three models (ERMIN 2, RESRAD-BUILD, and METRO-K) used an element based approach, in which radiation doses are estimated by adding the contributions to dose rate from various surfaces (e.g. roof, walls, road, etc.). ERMIN 2 and METRO-K use surface specific retention functions to model radionuclide behaviour on the surfaces; RESRAD-BUILD models the dynamic transfer of radionuclides between surface compartments.

A probabilistic approach was used with ERMIN 2, TINT-UNSCEAR, DPRO, and METRO-K, while a deterministic approach was used with RESRAD-BUILD [14]. Probabilistic approaches considered distributions for the surface contamination levels and for the time spent in various places.

TABLE 17. SUMMARY OF MODELS AND MODELLING APPROACHES USED FOR THE FUKUSHIMA DOSE ASSESSMENT EXERCISE (reproduced from Ref. [14] with permission courtesy of Journal of Radiological Protection)

Model name	DPRO	ERMIN 2	RESRAD-BUILD	TINT-UNSCEAR	METRO-K
Participant and country	S. Takahara Japan	T. Charnock United Kingdom	C. Yu, S. Kamboj United States of America	K. Silva Thailand	J. E. Lee Republic of Korea
Purpose of model	To assess representative values of radiation dose for management of emergency and existing exposure situations.	To assess residual radiation doses in urban environments, and the impact of various cleanup or management options on those doses [44].	The RESRAD-BUILD computer code is a pathway analysis model to evaluate the radiation dose incurred by an individual who works or lives in a building contaminated with radioactive material [45].	A Microsoft Excel-based calculation sheet to assess the radiation dose for the representative person based on the measured ground concentration.	To assess the radiation dose to inhabitants in a radioactively contaminated Korean urban environment.
Starting point for modelling	The model input is the ground surface contamination level (Bq m^{-2}). Contributions from the other surfaces are not taken into account. However, indoor dose rate is calculated using a dose reduction factor (as described below). Since this factor is evaluated based on the actual measurements in Fukushima, contributions from other surfaces (e.g. house wall, roof) may be considered implicitly.	The principal input is deposition onto a lawn or grass surface at some distance from other urban surfaces (e.g. buildings, trees, roads). Other inputs include the weather conditions at the time of deposition (i.e. wet or dry), the type of urban environment being considered (selected from a library of typical environments), and the cleanup options being applied.	The initial deposition onto exterior walls, roof, outdoor grass and paved areas was assumed. The surface densities used on different surfaces in the code were estimated from the activity concentration data provided in the scenario description, corrected for weathering and decay.	The model uses the ground activity concentration as its input. It then calculates the dose to the representative person taking into account the composition of radionuclides deposited, migration of the radionuclides in soil, effective dose rate conversion coefficient, location factors (using different equations for paved surface, unpaved surface, wooden house, wooden fireproof house and concrete building), and occupancy factors (indoor versus outdoor; home versus workplace).	The starting point is the activity concentration in air (Bq m^{-3}). Five types of surface are considered: roof, outer wall, paved road, soil or lawn, and tree. Deposition is calculated using a deposition velocity in the case of no precipitation at the time of an accident. Deposition is calculated using a washout ratio in the case of precipitation on the date of an accident. Runoff is considered, if precipitation exceeds the critical amount of precipitation.
Indoor contamination	Dose from external exposure inside a house is assessed using dose reduction factors for a typical Japanese house [34].	ERMIN 2 applies a set of ratios that relate the amount of indoor deposition of a radionuclide to the outdoor deposition. A ratio is selected depending on the deposition conditions and particle group of the radionuclide. The ratios have been calculated externally using an expression that relates building dimensions, filtration factor, air exchange rate, and indoor deposition rate [46].	Indoor floor contamination is assumed to be some fractions of the outside contamination (i.e. 10%, 5%, 3%, and 1%) [47].	Not included. Only doses from the outdoor contaminated area to the representative person inside a wooden house, a wooden fireproof house and a concrete building are estimated.	Indoor contamination is not considered in the current model.

TABLE 17 (cont.)

Model name	DPRO	ERMIN 2	RESRAD-BUILD	UNSCEAR-TINT	METRO-K
Weathering	An attenuation function is used for modelling the effects of weathering. This function was developed by Kinase et al. [48] based on measurements after the Fukushima accident. In addition, this function is represented as a two exponent model; one exponent describes the distribution process and the other describes the elimination process in the local environment. Kinase et al. [48] use eight categories for the local environment: water, urban, paddy, crop, grass, deciduous forest, evergreen forest and bare surface.	ERMIN 2 explicitly models weathering from and retention on different urban surfaces. For most surfaces, it uses empirical functions in the form of one or two term exponential expressions. Migration down the soil column is modelled using a convective dispersive model [49]. For deciduous trees, an instantaneous leaf fall event is assumed, while it is assumed that coniferous trees shed needles continuously.	The weathering correction factors adopted in the RESRAD-RDD code [47] (originally taken from Ref. [50]) were used. Because of the nature of the roofs in a typical Japanese house, the weathering for roof was assumed to be the same as for a paved area. The weathering correction for soil was applied to grass and paved areas outside.	Simple exponential equations are used to calculate the dose from paved and unpaved surfaces [40]. TINT-UNSCEAR assessments were performed by using the two parameters for paved and unpaved surfaces based on the ratio between them shown in Fig. 116 above.	Weathering is modelled using Gale's equation, which is distinguished by two exponential terms [51]. Constants and weathering half-lives in the model are a function of surface type.
Retention of radionuclides, outdoors	Not included in the model	See weathering above; the external surfaces modelled include paved, grass or plant leaves, walls, roofs, tree surfaces (trunk and limbs), leaves and needles, and soil column.	Weathering correction factors were used.	Simple exponential equations were used to calculate the dose associated with paved and unpaved surfaces.	Retention is considered in terms of the retained fraction in runoff water, which is a function of surface type and radionuclide.
Retention of radionuclides, indoors	Not included in the model	See weathering above; the indoor surfaces are represented as one simplified indoor surface.	Weathering correction factors were used.	Not included in the model	Indoor contamination is not considered in the current model.
Mobile fractions (by surface) and half-lives	Not included in the model	See Ref. [52].	See Ref. [47].	Not included in the model	10% of initial deposition is considered as the mobile fraction for dry deposition. For this fraction, the daily fixation rate is 70%. For wet deposition, 100% of initial deposition is the non-mobile fraction [53].

TABLE 17 (cont.)

Model name	DPRO	ERMIN 2	RESRAD-BUILD	UNSCEAR-TINT	METRO-K
Most important surfaces with respect to contribution to external dose	Dependent on the deposited contamination, and the occupancy factor.	Dependent on the urban environment, the radionuclide mix, the time after deposition and the deposition conditions. Very generally the grass/soil surface is expected to be the most important surface for most times (even for this exercise with smaller lawns around the houses). Followed by trees (until leaf fall), roofs and roads. Walls are expected to be the least important due to low initial deposition. Interior surfaces may be important in highly shielded environments.	Dependent on the deposited contamination and building configuration.	The most important surfaces depend on the occupancy factors (factors indicating the time the representative person spent near each surface). In this case, the wooden house has the highest contribution since people spend most time at home in both cases (indoor workers and outdoor workers). However, for the same occupancy factor, the unpaved surface contributes the highest external doses.	The external dose depends on the typical Korean surrounding environment (prefabricated house, detached house, terraced house, business building, apartment) and location of receptor.
Probabilistic calculations	Yes	Operational version: No Research version: Yes	No	Yes	Yes
External dose coefficients	From Ref. [54].	ERMIN 2 has a library of urban environments. For each environment, radionuclide and urban surface, the library contains dose rates from the surface to locations indoors and outdoors for a unit deposition. The library has been compiled from existing studies that used Monte Carlo particle transport techniques to calculate dose rates in different urban environments [55, 56].	The RESRAD-BUILD computer code uses external dose coefficients from Federal Guidance Report No. 12 [57].	From Refs [40, 58].	Air absorbed dose is calculated using a pre-calculated kerma value which is a function of the radionuclide energy and location of the receptor [56]. The external dose is then calculated using a dose conversion factor (Sv per Gy) and a correction factor to account for surface roughness.
mGy to mSv conversion factor	The dose is estimated directly from the deposition, based on the effective dose rate conversion coefficient (e(dep,m) (nSv h^{-1} per kBq m^{-2}))	Approximated using ICRP conversion factors [59].	The code uses external (and internal) dose coefficients in dose calculations, and mGy to mSv conversion is included in the dose coefficients.	The dose is estimated directly from the deposition, based on the effective dose rate conversion coefficient (e(dep,m) (nSv h^{-1} per kBq m^{-2}))	To convert from air absorbed dose to external dose to a receptor, 0.8 and 0.7 mSv/mGy are applied for outdoor and indoor residents, respectively.
Remediation countermeasures (Calculations in this exercise were performed without consideration of countermeasures)	Not included	ERMIN 2 has implemented most of the options in the EURANOS Inhabited Areas Handbook [60].	Not included, but various countermeasures can be simulated by adjusting input parameters for various scenarios.	Not included	Not included

7.4. ANALYSIS OF MODELLING RESULTS

Results of the dose assessment exercise are shown in Figs 116 and 117 reproduced from [14]. Figure 116 shows the measured and predicted distributions (from four models) of external dose (annual effective dose) for indoor and outdoor workers, and the predicted deterministic dose from the fifth model. Figure 117 compares the arithmetic mean and selected percentiles from the measurements and the models.

7.4.1. Doses to indoor workers

Predictions of doses to indoor workers from ERMIN 2, TINT-UNSCEAR, and DPRO were very close to each other and to the measurements (Fig. 116). The DPRO model was developed from this set of measurements [34], with the capability of accounting for various factors in the dose assessment, such as deposition conditions and housing types in Fukushima City; thus, predictions from DPRO agree well with the actual measurements of doses.

Selected percentiles and the arithmetic means of measurements and model predictions are compared in Fig. 117. For indoor workers, differences between the predicted and measured values for the 5th, 50th, and 95th percentiles and the arithmetic means were within a factor of 1.2 for ERMIN 2, TINT-UNSCEAR, and DPRO, while the difference between the predicted and measured 5th percentiles was a factor of 2 for METRO-K.

In general, the DPRO predictions agreed well (less than a factor of 1.2) with the actual measurements between the 5th and 95th percentiles, but the predicted maximum was a factor of 4.5 higher than the measured maximum value [14]. The maximum value of a probabilistic model result depends strongly on the truncation level (the upper limit for the generation of random numbers in numerical calculations) used for normal and lognormal distributions.

RESRAD-BUILD used a deterministic approach, based on use of the arithmetic mean of the time spent in various places. The predicted dose for the indoor worker from RESRAD-BUILD agreed well with the arithmetic mean of other models assessed stochastically (within a factor of 1.4 of the other model predictions; Fig. 117). This value also reproduced the arithmetic mean of the measured values quite well (within a factor of approximately 1.1).

7.4.2. Doses to outdoor workers

The differences among model predictions for outdoor workers were larger than those for indoor workers (Fig. 116). For example, the predicted arithmetic means ranged from 1.2 to 1.9 mSv/y (Fig. 117). The lowest predicted value of the arithmetic mean for outdoor workers was from TINT-UNSCEAR; this model used a location factor of 0.75 for time spent by an outdoor worker in an outdoor workplace. The other location based model, DPRO, assumed a location factor of 1.0, and its prediction agreed well with the measurement.

Other possible explanations for the larger differences among model predictions for outdoor workers compared with indoor workers include differences in other parameters related to outdoor work. For example, surface contamination levels for agricultural land are different from those for residential areas, which is important for estimation of doses to agricultural workers. Another difference could be the distribution form used for occupancy times for outdoor versus indoor workers.

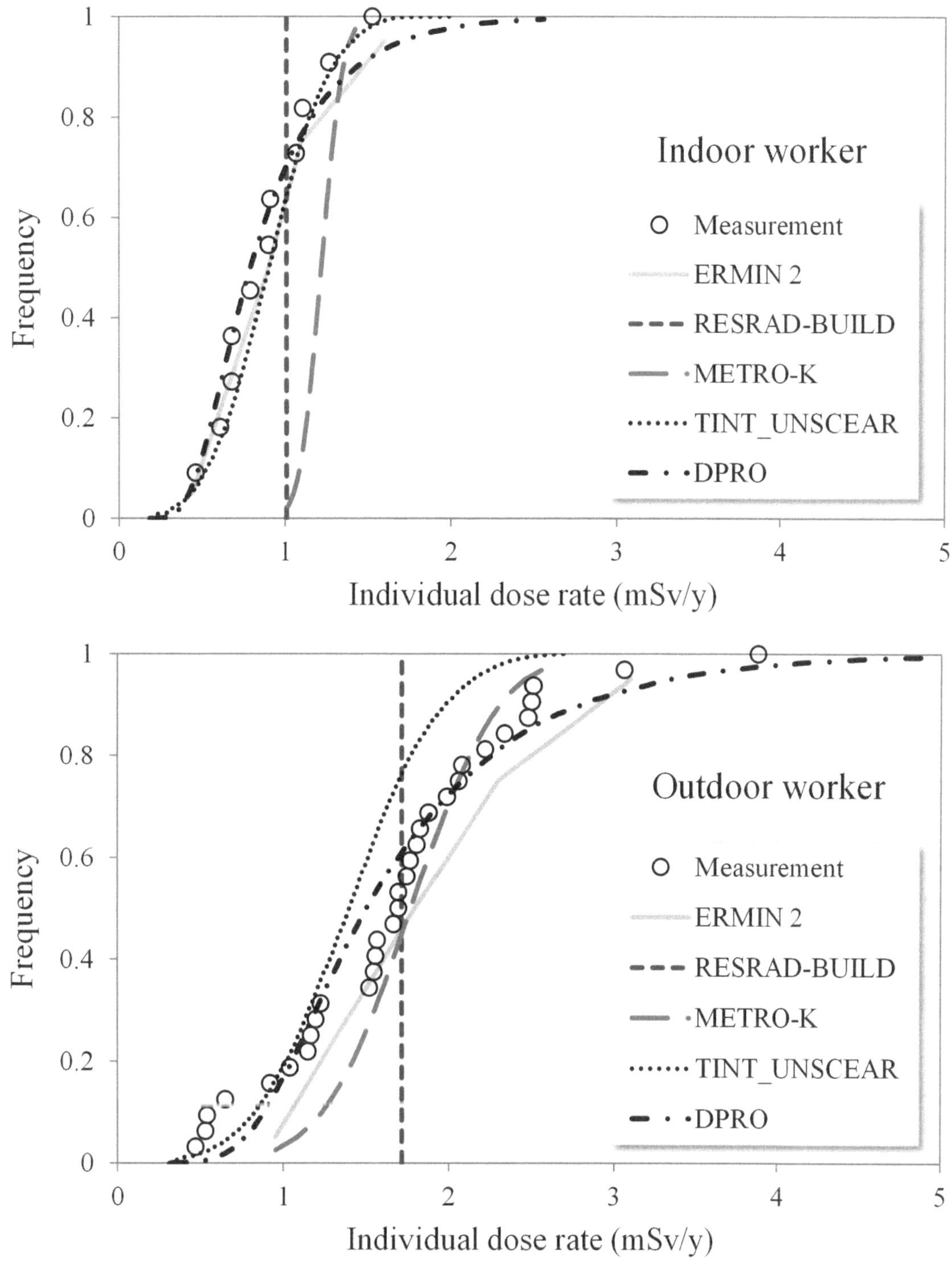

FIG. 116. Assessment results for the distribution of annual effective dose for indoor workers (top) and outdoor workers (bottom). The assessments were performed for the period from February 2012 to January 2013 (figure reproduced from Ref. [14] with permission courtesy of Journal of Radiological Protection).

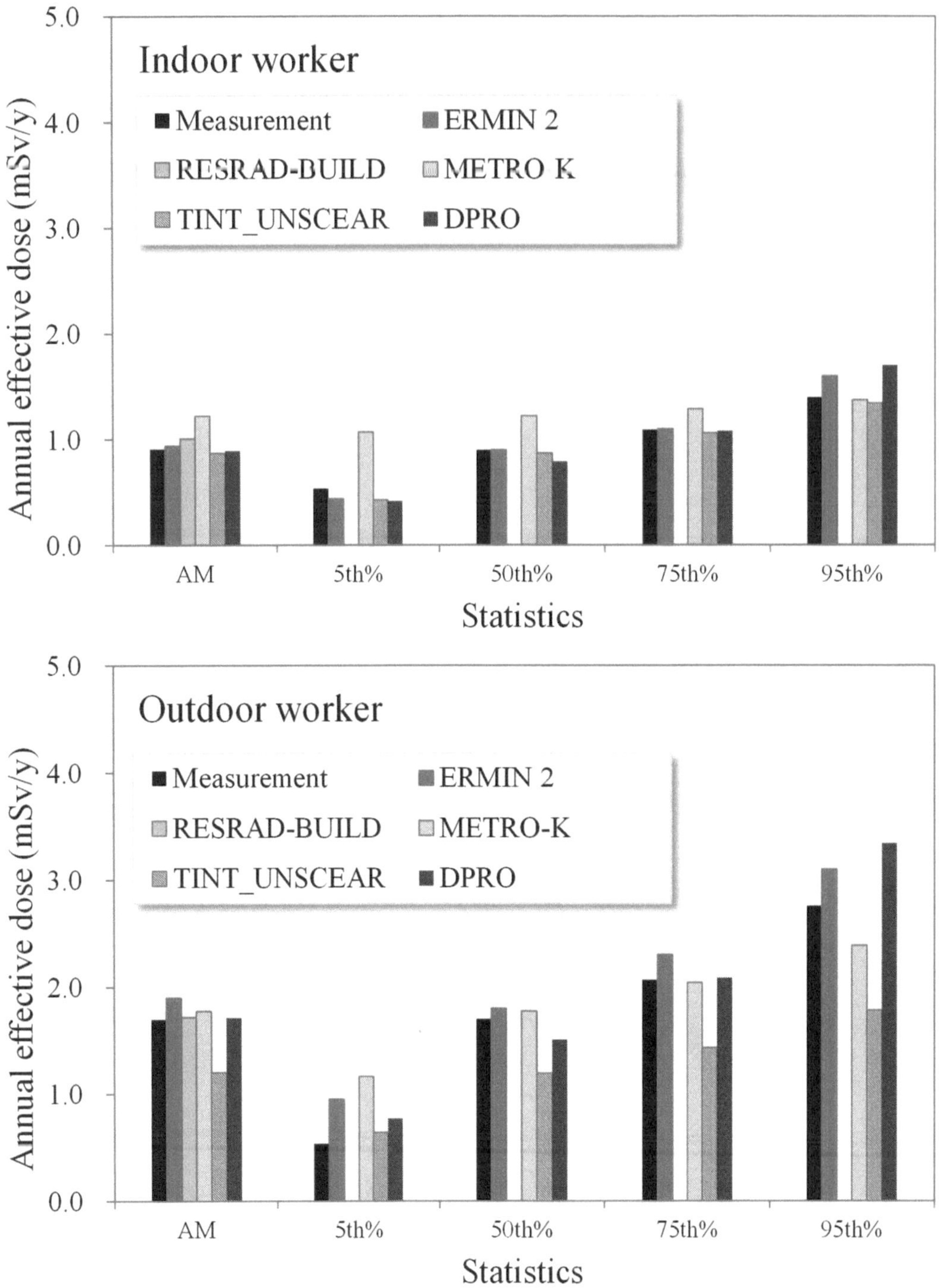

FIG. 117. Comparison of statistics (predicted and measured) for the annual effective dose for indoor workers (top) and outdoor workers (bottom). (AM = arithmetic mean; also shown are the 5th, 50th, 75th, and 95th percentiles). The assessments were performed for the period from February 2012 to January 2013 figure reproduced from Ref. [14] with permission courtesy of Journal of Radiological Protection).

7.4.3. Comparison of model types

Two probabilistic modelling approaches were used in this exercise, a location based approach (DPRO and TINT-UNSCEAR) and an element based approach (ERMIN 2 and METRO-K). The results shown in Fig. 116 indicate that either approach can provide a useful distribution of predicted doses, in good agreement with the measurements. However, given that some of the models obtain key information from the same references, the models might not be greatly different, at least in the context of this exercise.

There were some differences in the results among the three element based models (ERMIN2, METRO-K and RESRAD-BUILD). As described in Table 17, the three element based models use different approaches for indoor contamination (METRO-K did not include indoor contamination), weathering, and retention of radionuclides outdoors and indoors. Differences in the results are therefore due not only to the handling of input data for the surface contamination, but also to the differences in model components. For the two location based models (DPRO and TINT-UNSCEAR), the primary difference between them was whether or not a location factor for the outdoor worker was used [34], as described above.

7.4.4. Comparison of approaches for prediction of doses to the representative person

The representative person is a concept used in the context of radiation protection for members of the public by the ICRP [61] and the IAEA Safety Standards [35]. ICRP Publication 101 recommends that "the representative person should be defined such that the probability is less than about 5% that a person drawn at random from the population will receive a greater dose" [61]. It also recommends that doses to such individuals be assessed using both probabilistic and deterministic approaches.

Table 18 shows the predicted 95th percentile values from ERMIN 2, METRO-K, TINT-UNSCEAR and DPRO for doses to the representative person for indoor and outdoor workers in Fukushima City in 2012. For a probabilistic approach, the predicted 95th percentiles of the doses to the representative person agreed well with the 95th percentile of the measured values. However, in some cases the predicted values were not conservative, that is, they were lower than the measured 95th percentile values. The model predictions were based on the assumption of representative houses and their surrounding environment; however, this assumption may have resulted in results that were not conservative, due to the inherent uncertainty and variability associated with this assumption.

When the results from probabilistic and deterministic calculations obtained using the same model were compared (METRO-K, TINT-UNSCEAR and DPRO; Table 18), the results from the deterministic calculation were always higher than those from the probabilistic calculation [14]. This is consistent with previous studies, which found overly conservative results due to use of conservative values for most input parameters [62]. Use of both the arithmetic mean of the surface deposition (greater than the geometric mean for the case of a lognormal distribution such as in this case [33]) together with the 95th percentile of the time spent outdoors (based on Ref. [61]) apparently resulted in excess conservatism in the present exercise [14].

TABLE 18. MEASURED AND PREDICTED DOSE TO THE REPRESENTATIVE PERSON FOR INDOOR AND OUTDOOR WORKERS IN FUKUSHIMA CITY IN 2012[a]

Model	Annual effective dose to the representative person (mSv/y)		
	Measurements[b]	Probabilistic approach[b]	Deterministic approach[c]
Indoor workers			
Measurements	1.40		
ERMIN 2		1.60	—
METRO-K		1.39	1.50
TINT-UNSCEAR		1.34	1.56
DPRO		1.70	1.77
Outdoor workers			
Measurements	2.76		
ERMIN 2		3.1	—
METRO-K		2.46	3.37
TINT-UNSCEAR		1.78	2.63
DPRO		3.34	3.76

[a] Table reproduced from Ref. [14] with permission courtesy of Journal of Radiological Protection).[b] Dose to the representative person was determined as the 95th percentile in the distributions of the measurements or the modelled values. Values of the 95th percentile were taken from those shown in Fig. 117.

[c] The deterministic approach used the arithmetic mean of the surface deposition and the 95th percentile of time spent outdoors.

7.5. FINDINGS FROM THE FUKUSHIMA EXERCISE

Two types of models, location-based and element-based, were used in the Fukushima exercise [14]. The distributions of measured doses were reproduced by both types of models, although the predicted ranges sometimes exceeded the observed ranges. Predicted arithmetic means agreed well with the arithmetic means of the measurements, both for the probabilistic models and the one deterministic model (RESRAD-BUILD). In general, it was more difficult to reproduce measurements for outdoor workers than indoor workers, with greater differences among model predictions for the outdoor workers.

When the same model was used for both probabilistic and deterministic results, the deterministic approach consistently gave higher results. This may result from use of conservative values for many or most parameters in the deterministic approach, rather than consideration of uncertainty for each individual parameter as in a probabilistic approach.

Comparison of the 95th percentile values of the measured and predicted dose distributions showed that in some cases, the predicted 95th percentile values would not have been adequately conservative (protective) as considered would be the case in ICRP's guidance. In other words, the dose to the representative person would have been underestimated. For this particular situation, the uncertainty of important parameters (e.g. weathering and retention) and the variability in size and shape of houses would need further attention.

The first modelling exercise based on Fukushima data examined prediction of doses to specified reference individuals [11]. The present exercise examined prediction of the distribution of doses within a population, as well as evaluating the concept of the representative person. The modelling exercise could be extended in the future to use the models to predict the effectiveness of various types of remediation option, compare possible remediation strategies to predict the distribution of doses within a population.

8. CONCLUSIONS

Most of the modelling exercises described in this publication involved atmospheric dispersion situations, either short range, relevant to releases within an urban area (Sections 2 and 3) or mid-range, relevant to releases from larger facilities that could have an impact on downwind urban areas (Sections 4–6). Several general conclusions can be drawn from these modelling exercises:

- Meteorological data, ideally, need to be obtained as close as possible to a dispersion point, or be adjusted for spatial and time differences as needed;
- Time dependent endpoints (e.g. peak concentrations) at specific locations can be difficult to model; making allowance for spatial and temporal error can improve the model predictions;
- The effects of buildings or complex terrain on prediction of downwind dispersion can be difficult to model but need to be considered;
- Use of weather forecasts instead of meteorological measurements still presents a significant challenge.

An additional modelling exercise (Section 7) started with information on the initial contamination of an area, together with habit data for the population, to estimate distributions of doses within the population. This exercise also compared probabilistic and deterministic approaches for dose estimation, as well as the concept of the representative person. Further use of this dataset would be to undertake an exercise to the consider various remediation strategies and their expected effect on the distribution of doses within a population.

As has been demonstrated in these and the previous modelling exercises, intercomparison of predictions from several models and participants, and comparison of model predictions with measurements when available, can lead to improved understanding of the modelling process for given types of situations, as well as to improved model performance.

APPENDIX. DESCRIPTIONS OF MODELS

This Appendix includes more detailed descriptions of the application of the models used for the modelling exercises included in this publication, as listed below:

- Boletice modelling exercise (see Section 2): Sections I.1–I.3;
- Šoštanj modelling exercise (see Section 4): Section I.4;
- Chalk River exercise (see Section 6): Sections I.5–I.7;
- Fukushima modelling exercise (see Section 7): Sections I.8–I.10.

In addition, the description includes a brief summary of the model used for the modelling exercise.

Other models used in the modelling exercises covered in this publication are described in the EMRAS II report [2] and in the MODARIA I report [11].

A.1. DESCRIPTION OF HOTSPOT 3.0.3 (BOLETICE EXERCISE)

The HotSpot 3.0.3 code was used for the Boletice modelling exercise by Thomas Charnock of the UK Health Security Agency in the United Kingdom.

A.1.1. Introduction

The HotSpot 3.0.3 program, developed by the Lawrence Livermore National Laboratory, was used for the Boletice exercise. HotSpot 3.0.3 contains several models for different situations; for this modelling exercise the general explosion model was used.

The developers of HotSpot are quite clear about the purpose and scope of the software. The following text is extracted from the HotSpot 3.0 user guide [63]:

> "The HotSpot Health Physics Codes, or HotSpot program, provides a first-order approximation of the radiation effects associated with the atmospheric release of radioactive materials. The HotSpot program was created to equip emergency response personnel and planners with a fast, field-portable set of software tools for evaluating incidents involving radioactive material. The software is also used for safety-analysis of facilities handling radioactive material. This program is designed for short range (less than 10 km), and short term (less than a few hours) predictions."

Full details of the model can be found in the HotSpot user guide [63]. In summary, the general explosion model applies the well-established straight line Gaussian plume formulation to a set of virtual source terms that are generated using a simple formulation to represent the vertical distribution of the activity in the column immediately following the explosion. Deposition onto the ground surface is modelled using dry deposition velocities and a rain out coefficient.

HotSpot considers partitioning of the activity within the column and subsequent atmospheric dispersion within a three dimensional frame of reference and deposition onto a two dimensional frame of reference.

The user provides the total activity of each radionuclide and the fraction of activity that is airborne. HotSpot requires the activity to be partitioned into two particle size groups — respirable and non-respirable — and the user needs to specify the fraction of activity in each group and the respective dry deposition velocities. The terms 'respirable' and 'non-respirable' can be misleading in the Boletice exercise context, which requires no assessment of inhalation dose. For this exercise, they merely represent two particle groups with different deposition velocities.

Additionally, the user needs to provide a meteorological description that includes wind speed at 2 m height, wind direction and stability category and choose a terrain type from either 'standard' or 'urban'.

The user provides an amount of explosive in lb TNT equivalent. Users can adjust the way the activity is distributed in the initial column, or they can accept the default parameters.

There are also various options concerned with estimating dose, but these were not used in this modelling exercise.

HotSpot general explosion model endpoints include:

- Time integrated activity concentration in air;
- Total deposition to the ground surface;
- Total effective dose equivalent from internal exposure to inhaled radionuclides;
- Dose rate from radionuclides deposited on the ground;
- Plume arrival time.

Key assumptions, modelling approaches and parameter values are in the HotSpot User's Guide [63]. Hotspot does not handle uncertainties in model parameters apart from an option to include different meteorological conditions.

A.1.2. Application to the Boletice exercise

An earlier but very similar version of HotSpot was used for a previous modelling exercise undertaken under the EMRAS II programme [2]. Under that programme, HotSpot was calibrated to the results of two 'open' experiments provided for that purpose. The parameters included in the calibration exercise were the partitioning into the two size groups, the deposition velocities of those groups, and the wind speed. The parameters were adjusted by trial and error to fit the predicted results to the reported deposition along the plume centre line from 10 m to 50 m. (HotSpot does not give results closer than 10 m, and the test results did not extend beyond 50 m).

In the EMRAS II exercise [2], the calibrated HotSpot model was then used to predict the results of several 'blind' experiments. The results were mixed, but generally not good. There are several reasons for this, but it is likely that both direct ballistic particles and gravitational settling make a significant contribution to deposition in the first 50 m, and neither process was included in the HotSpot model. (Hotspot does have a model for ballistic particles, but this was not used.) It was concluded that the calibration process adopted, amounted to little more than a curve fitting exercise.

For the exercise in MODARIA II, minimal calibration was performed, and default parameters were used whenever possible. The defaults used are either those provided in the HotSpot interface or those that would be adopted by the UK Health Security Agency (UKHSA) if faced with such an incident and with no additional information. An initial set of results was generated and following a working group meeting in which the results of several models were compared with some of the experimental data, the wind speed and wind direction were modified. The inputs used are given in Table 19.

A.1.3. Results

The purpose of the exercise was to predict deposition measured within 50 m of the detonation point. However, for UKHSA, the contamination in the immediate vicinity of the bomb is less important than that further away. In an actual incident, a cordon would be established at an appropriate distance from the site of detonation, for example, at 400 m. The public would be excluded from this area, and hence there would be no immediate public health concerns to be addressed. Furthermore, the area would be treated as a crime scene and subject to forensic examination. Ultimately it would be cleaned and intensively monitored before the cordon was lifted. However, beyond the cordon, UKHSA would be expected to provide advice as to appropriate actions for public protection and on a short time scale. Therefore, the results provided below go beyond 50 m, and whilst there are no experimental data to compare them to, there is the potential for model–model comparison (Figs 118 and 119).

TABLE 19. INPUT INFORMATION USED WITH HOTSPOT (UKHSA) IN THE BOLETICE EXERCISE

Parameter	Final run (initial run)	Notes
Stability category	D	UKHSA daytime default value
Wind direction (from)	286 (250)	Wind speed and direction both fluctuated significantly as shown by the graphs provided in the scenario description. The initial values used were selected from the dataset at the exact time of the detonation. However, the meteorological data were taken from a point about 110 m from the explosion, and there was evidence of a short lag between the weather recorded and the weather at the point of detonation. The new values were chosen from the data to account for the lag; they are well within the range of fluctuation, and the direction matches the observed plume deposition pattern.
Wind speed (m/s)	1.2 (4)	
Release height (m)	0	From scenario description
Explosive (TNT equivalent)	0.52 lbs	From scenario description, converted to pounds using a factor provided by a working group member. HotSpot predicts an initial column height of 17 m, which compares well with that observed and that assumed by other modellers.
Respirable fraction (division of release into two particle size groups)	0.2	HotSpot default
Respirable fraction deposition velocity (cm/s)	0.3	HotSpot default value
Non-respirable fraction deposition velocity (cm/s)	40	HotSpot default value

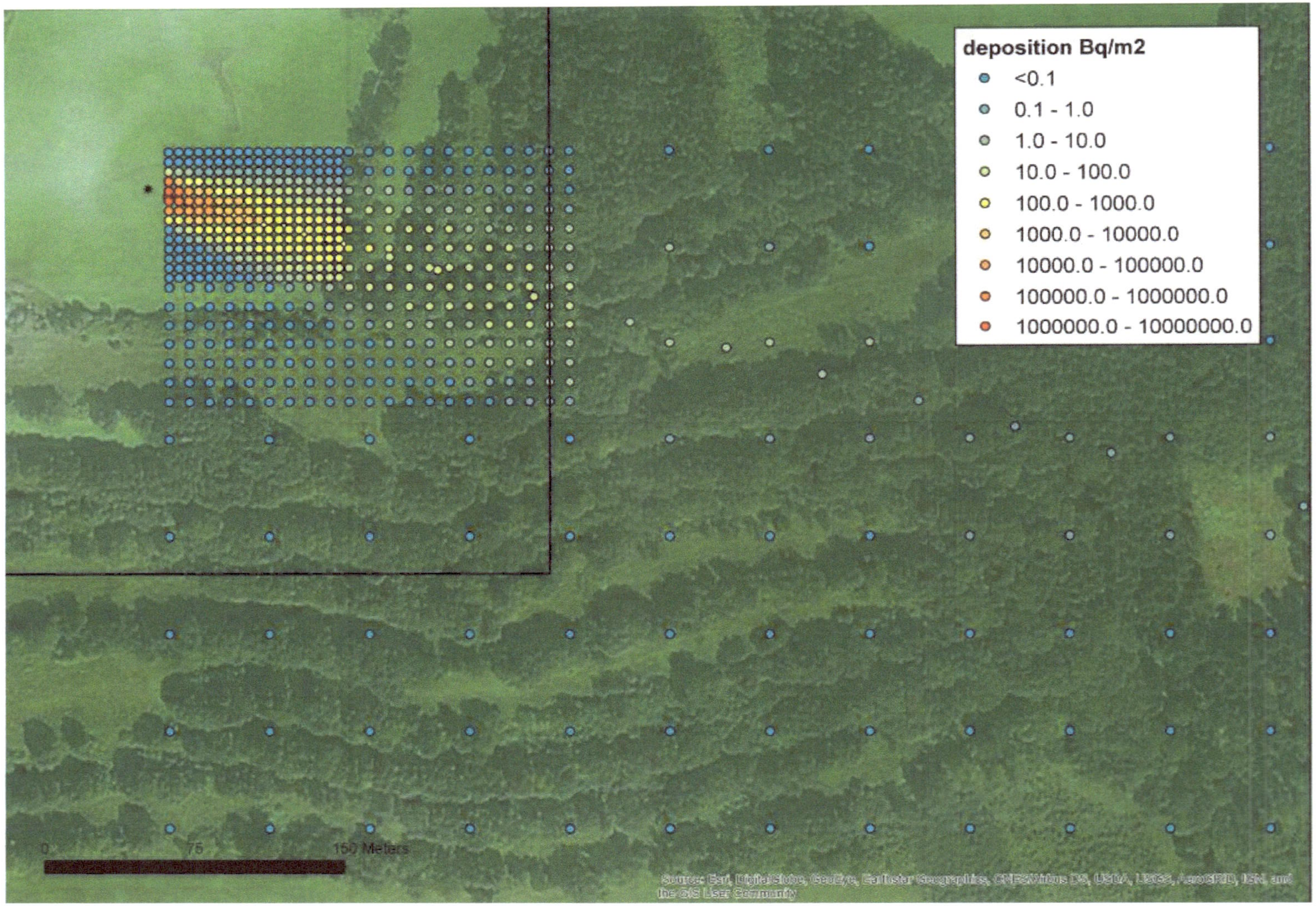

FIG. 118. The predicted deposition at the points of the nested grid.

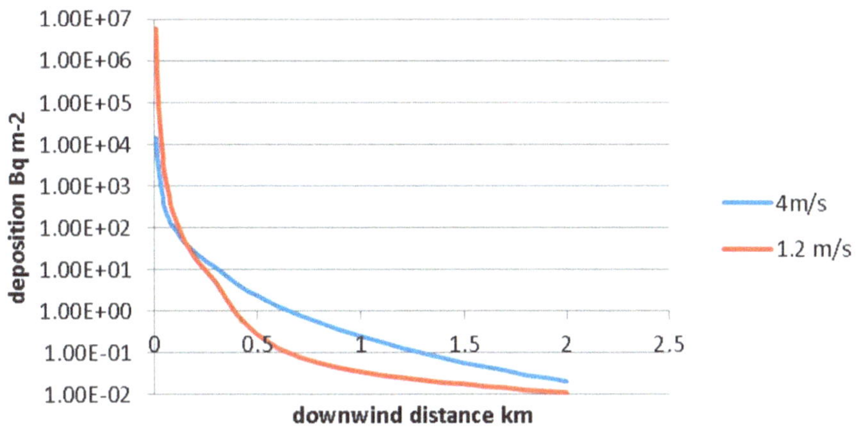

(a)

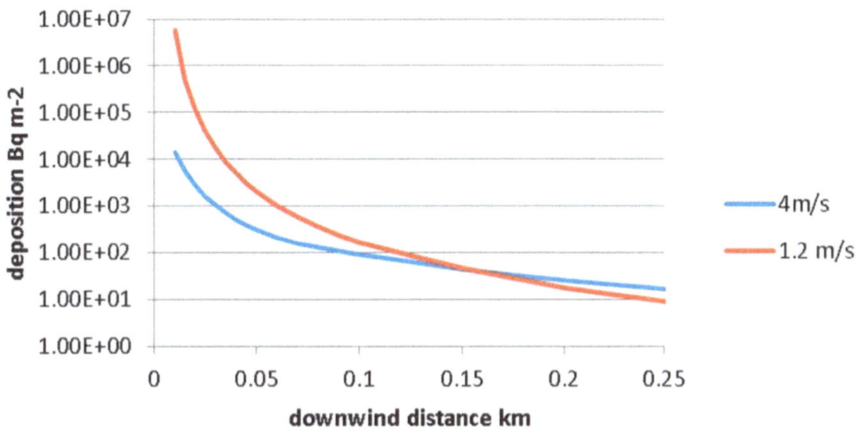

(b)

FIG. 119. The predicted deposition (Bq/m^2) along the plume centre line out to (a) 2 km and (b) 250 m.

A.1.4. Technical note

Generally, HotSpot is run as an interactive tool and can provide predictions at up to 20 locations further than 10 m from the point of release. However, it is possible to run it in a batch mode (called 'automatic mode'). It is by running HotSpot in automatic mode repeatedly, using a Python script, that the nest grid arrangement of locations used in this exercise was achieved.

To run in automatic mode, the default input file 'current.hot' needs to be edited before HotSpot is started. According to the user guide [63], the item 'SystemName' has to be changed as follows:

```
SystemName = Hotspot Automatic Table
```

However, this syntax will give results only along the plume centre line. It is an undocumented feature that the command to give results at locations specified as coordinates is:

```
SystemName = Hotspot Automatic Table Compass
```

A.2. DESCRIPTION OF HOTSPOT 3.1 (BOLETICE EXERCISE)

The HotSpot 3.1 code was used for the Boletice modelling exercise by Francesco Mancini of Sogin in Italy.

A.2.1. Introduction

HotSpot 3.1 uses Gaussian models to describe the atmospheric dispersion. The model limits the maximum downwind distance to 200 km and the minimum distance to 0.01 km, as it is generally as extrapolation of the σ_y and σ_z data below a distance of 10 meters is generally not advisable [63].

In this code, it is assumed that the target individual remains at the same downwind location (x, y, z) throughout the passage of the plume.

HotSpot includes atmospheric dispersion models for:

— General plume;
— A plutonium explosion (non-nuclear), fire, and resuspension;
— A uranium explosion (non-nuclear), and fire;
— A tritium release.

These models estimate the short range, downwind radiological impact following the release of radioactive material resulting from a short term release (less than a few hours), explosive release, fuel fire, or an area contamination event.

A.2.2. Application to the Boletice exercise

A.2.2.1. Key assumptions

For this exercise the Explosion (Non-nuclear) Model is used. The Explosion (Non-nuclear) release is partitioned in five separate area sources to model the initial distribution of material. Each of the 5 area sources [h(1) to h(5)] is represented by two separate upwind virtual source terms. These two virtual source terms are associated with either the horizontal (crosswind) or the vertical components of the area source. Table 20 summarizes the main data used for the general explosion model.

A.2.2.2. Parameter values

The respirable fraction was assumed to be 0.6, with a deposition velocity of 0.3 cm s^{-1}. For the non-respirable fraction (0.4), a deposition velocity of 10 cm s^{-1} was used.

A.2.2.3. Meteorological data

Tables 21–22 provide the measured wind direction and wind velocity, respectively, during the first minute after the explosion. For each, the median values for the first minute after the explosion were used. A value of 45 degrees was added to the wind direction to account for the real diffusion of the plume as observed from the video of the explosion.

TABLE 20. SUMMARY OF RELEASE DATA

Item	Type or Value
Model type	General explosion
Amount of ^{140}La	0.713 GBq
Release height	0 m

TABLE 21. METEOROLOGICAL DATA – WIND DIRECTION

Time	Wind direction (deg)	Time	Wind direction (deg)	Time	Wind direction (deg)	Time	Wind direction (deg)	Time	Wind direction (deg)	Time	Wind direction (deg)
17:32:00	230	17:32:10	250	17:32:20	243	17:32:30	253	17:32:40	248	17:32:50	245
17:32:01	238	17:32:11	255	17:32:21	246	17:32:31	250	17:32:41	245	17:32:51	248
17:32:02	242	17:32:12	247	17:32:22	247	17:32:32	260	17:32:42	237	17:32:52	251
17:32:03	245	17:32:13	238	17:32:23	247	17:32:33	259	17:32:43	239	17:32:53	255
17:32:04	245	17:32:14	236	17:32:24	248	17:32:34	254	17:32:44	239	17:32:54	257
17:32:05	248	17:32:15	236	17:32:25	243	17:32:35	249	17:32:45	232	17:32:55	253
17:32:06	258	17:32:16	232	17:32:26	252	17:32:36	249	17:32:46	236	17:32:56	253
17:32:07	247	17:32:17	230	17:32:27	260	17:32:37	248	17:32:47	238	17:32:57	255
17:32:08	248	17:32:18	236	17:32:28	265	17:32:38	248	17:32:48	243	17:32:58	247
17:32:09	250	17:32:19	241	17:32:29	258	17:32:39	250	17:32:49	239	17:32:59	238

TABLE 22. METEOROLOGICAL DATA – WIND SPEED

Time	Speed (m/s)	Time	Speed (m/s)	Time	Speed (m/s)	Time	Speed (m/s)	Time	Speed (m/s)	Time	Speed (m/s)
17:32:00	0.79	17:32:10	0.79	17:32:20	1.36	17:32:30	1.42	17:32:40	1.77	17:32:50	1.48
17:32:01	0.83	17:32:11	0.68	17:32:21	1.36	17:32:31	1.53	17:32:41	1.96	17:32:51	1.82
17:32:02	0.86	17:32:12	0.65	17:32:22	1.34	17:32:32	1.61	17:32:42	1.76	17:32:52	2.04
17:32:03	0.94	17:32:13	0.77	17:32:23	1.29	17:32:33	1.71	17:32:43	1.56	17:32:53	1.91
17:32:04	0.99	17:32:14	0.95	17:32:24	1.33	17:32:34	1.74	17:32:44	1.6	17:32:54	1.74
17:32:05	0.74	17:32:15	1.09	17:32:25	1.29	17:32:35	1.56	17:32:45	1.6	17:32:55	1.52
17:32:06	0.69	17:32:16	1.06	17:32:26	1.25	17:32:36	1.59	17:32:46	1.33	17:32:56	1.45
17:32:07	0.82	17:32:17	1.04	17:32:27	1.24	17:32:37	1.91	17:32:47	1.66	17:32:57	1.64
17:32:08	0.82	17:32:18	1.09	17:32:28	1.11	17:32:38	1.96	17:32:48	1.49	17:32:58	1.69
17:32:09	0.88	17:32:19	1.23	17:32:29	0.92	17:32:39	1.77	17:32:49	1.37	17:32:59	1.58

TABLE 23. INPUT INFORMATION USED WITH HOTSPOT 3.1 IN THE BOLETICE EXERCISE

Parameter	
Explosion time	17:32:00
Wind speed (m/s)	1.33
Wind direction (deg)	291
Stability category	B
TNT Equivalent (lb)	0.52
Source activity ^{140}La (GBq)	0.713
Respirable fraction	0.6
Deposition velocity of respirable fraction (cm/s)	0.3
Deposition velocity of non-respirable fraction (cm/s)	10

A.2.2.4. *TNT Equivalent*

Equation (1) given below was used to quantify the TNT-Equivalent of the explosive used (250 g of SEMTEX 1A) for the experiment:

$$250 \text{ g} \times (4980 \text{ kJ/kg}) / (4184 \text{ kJ/kg}) \times 0.8 = 238 \text{ g} = 0.52 \text{ lb} \quad (1)$$

where:

Mass of SEMTEX 1A = 250 g;
Explosion heat of SEMTEX 1A = 4980 kJ/kg;
Explosion heat of TNT = 4184 kJ/kg;
Relative work ability = 0.8.

A.2.2.5. *Summary of input information*

Table 23 summarizes the input information used with HotSpot 3.1 for the Boletice modelling exercise.

A.2.3. Results

Figure 120 shows the predicted deposition in the test area, including the predicted maximum deposition.

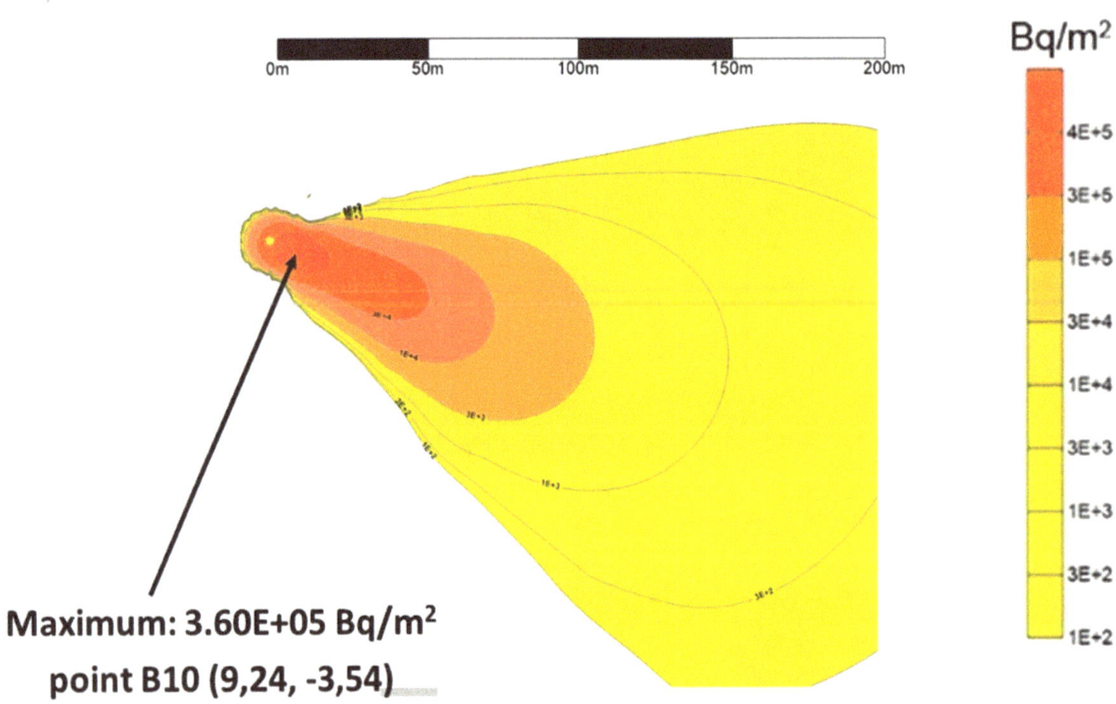

FIG. 120. Predicted deposition in the experimental area, showing the maximum deposition (Bq/m^2).

A.3. DESCRIPTION OF LASAIR 4.0.5 (BOLETICE EXERCISE)

The LASAIR 4.0.5 code was used for the Boletice modelling exercise by Francesco Mancini of Sogin in Italy.

A.3.1. Introduction

The code LASAIR (Lagrangian simulation of the dispersion and inhalation of radionuclides) has been developed by the Federal Office for Radiation Protection (Germany) to simulate atmospheric dispersion of radionuclides after an accidental release [64]. The model assists in such cases, or in cases of malevolent threats, with a model domain of approximately 20 km × 20 km and the finest grid size of 5 m × 5 m. The model is based on a well-accepted mathematical procedure (Lagrangian particle procedure), with a state of the art turbulence parameterization and a mass consistent diagnostic wind field model. The model provides for assessment of the radiation exposure after explosion, fire or short term momentum releases with special consideration of the radiation dose from inhalation, cloud shine and ground shine as well as activity concentration, deposition as a function of time or the ambient dose rate. The model is especially dedicated for operational use to assist police forces but can be applied as well for analysing the influence of building structures in order to provide information on the effects of instantaneous or long term emissions.

Within the Federal Office for Radiation Protection (BfS), the model LASAIR is used as a decision support tool in the context of malevolent attacks or the simulation of the dispersion of radionuclides in the close vicinity of nuclear installations.

A.3.2. Application to the Boletice exercise

A.3.2.1. Key assumptions

For this exercise the Explosion Model is used with a duration of release of 1 second. The explosion release is partitioned in a cloud with horizontal extension of 12 m and vertical extension of 20 m. The dimensions of the cloud depend on the mass of explosive. Table 24 summarizes the main data used for the general explosion model.

A.3.2.2. Parameter values

Table 25 summarizes the assumptions for the particle size distribution. A roughness length of 0.01 m was also assumed.

A.3.2.3. Meteorological data

For the meteorological data (wind velocity, wind direction), mean values were taken for each minute after the explosion, from 17:32 to 17:41 (Table 26). A value of 45 degrees was added to the wind direction to account for the real diffusion of the plume as observed from the video of the explosion.

TABLE 24. SUMMARY OF RELEASE DATA

Item	Type or value
Type of release	Explosion
Amount of explosive (TNT equivalent)	238 g
Horizontal extension of the cloud	12 m
Vertical extension of the cloud	20 m
Duration of the release	1 s
Amount of ^{140}La	0.713 GBq
Domain size	20 km × 20 km
Grid size	5 m × 5 m

TABLE 25. DUST COMPOSITION (PARTICLE SIZE DISTRIBUTION)

Particle size	%
< 2.5 μm	20
< 10 μm	60
< 50 μm	80
Total	100

TABLE 26. METEOROLOGICAL DATA

Time	Wind direction (deg)	Wind speed (m/s)	Stability class
17:32	291	1.33	B
17:33	306	0.86	B
17:34	326	1.13	B
17:35	322	1.24	B
17:36	280	1.12	B
17:37	297	0.97	B
17:38	325	0.65	B
17:39	305	0.98	B
17:40	313	0.77	B
17:41	297	0.58	B

A.3.2.4. TNT Equivalent

Equation (2) given below was used to quantify the TNT-Equivalent of the explosive used (250 g of SEMTEX 1A) for the experiment:

$$250 \text{ g} \times (4980 \text{ kJ/kg}) / (4184 \text{ kJ/kg}) \times 0.8 = 238 \text{ g} \quad (2)$$

where:

Mass of SEMTEX 1A = 250 g;
Explosion heat of SEMTEX 1A = 4980 kJ/kg;
Explosion heat of TNT = 4184 kJ/kg;
Relative work ability = 0.8.

A.3.3. Results

Figures 121 and 122 show the predicted deposition and predicted activity concentration in air, respectively, in the test area.

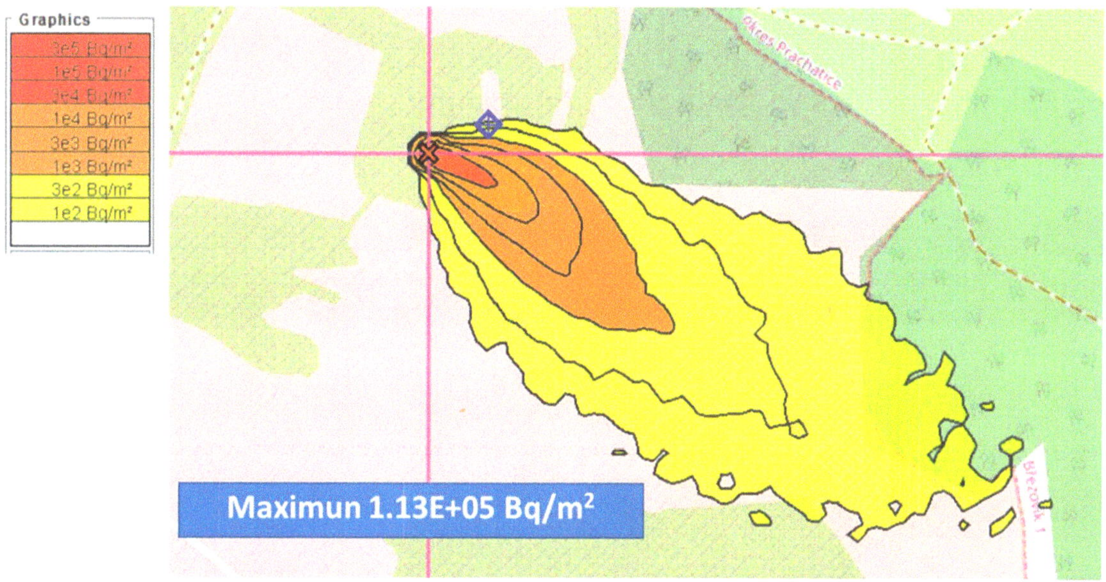

FIG. 121. Predicted deposition in the experiment area (Bq/m²).

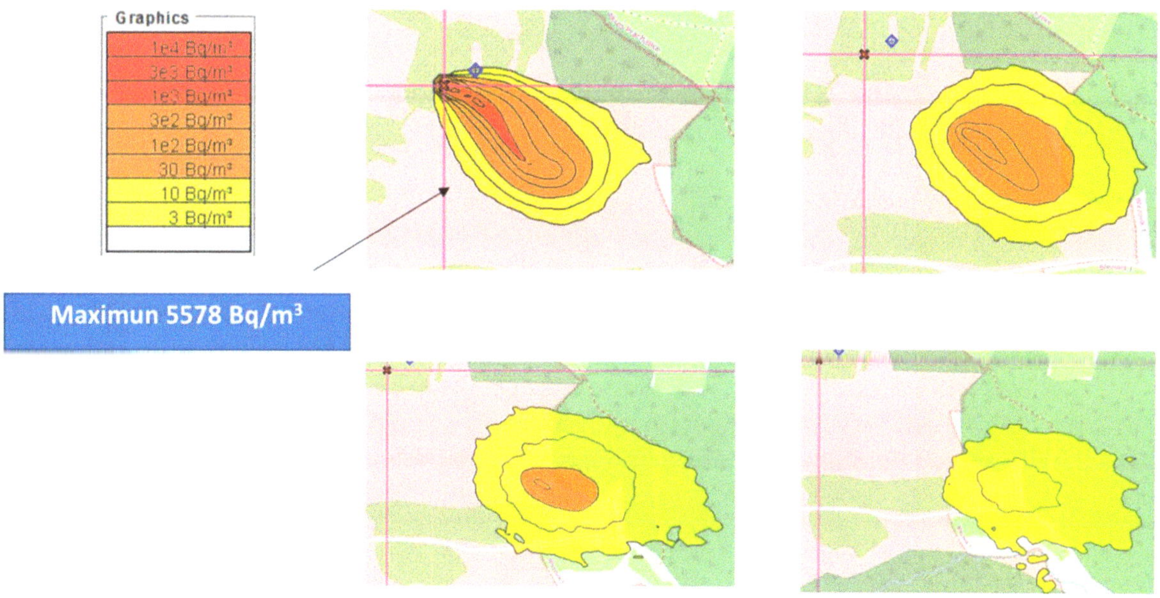

FIG. 122. Predicted activity concentration in air (Bq/m³).

A.4. DESCRIPTION OF ARTM (ŠOŠTANJ EXERCISE)

The ARTM code was used for the Šoštanj modelling exercise by Margit Pattantyús-Ábrahám of the Federal Office for Radiation Protection in Germany.

A.4.1. Introduction

The Atmospheric Radionuclide Transport Model (ARTM) [20] is an atmospheric dispersion model for regulatory purposes. The model is designed for simulating long term atmospheric dispersion of radionuclides from planned releases. The model consists of two subparts:

— Diagnostic wind field model TALdia, providing three dimensional wind and turbulence fields calculated from one measurement point of meteorological data. The turbulence field is optimized for one hour;
— Lagrangian particle tracking according to the results of the TALdia.

During the particle tracking, the radioactive decay, sedimentation, and wet and dry deposition are taken into account if necessary. The model permits the usage of time dependent weather parameters.

A.4.2. Application to the Šoštanj exercise

The orography data around the Šoštanj nuclear site was provided in the UTM 33T coordinate system with 100 m × 100 m resolution. As the modelled wind field revealed, the inhomogeneity of the terrain had to be considered (see Fig. 123).

A.4.2.1. Key assumptions

(1) The inhomogeneity of the terrain has to be taken into account.
(2) The diagnostic wind field model estimates the wind field from a point measurement. Here, the SODAR data was used at emission height (158 m) and at the approximate height of the neighbouring hills (376 m).
(3) The main effects on the wind field are assumed to be the detailed orography. Therefore, buildings were not taken into account for the simulation.
(4) The experiment does not deal with radioactive material, but with SO_2 emissions and its concentrations in air at specified monitoring points. ARTM deals with radioactive particles and emissions in terms of Bq/s. Sulfur-35 is used as the emitted molecule for the calculation. ARTM can only deal with S in aerosol and not gaseous form, therefore the smallest size (< 2.5 μm) is chosen for the emitted particles. one Bq of ^{35}S is taken as 1 μg SO_2, in order to obtain the same result quantity as the measurements (μg/m^3).

A.4.2.2. Parameter values

(1) Land cover is mostly mixed forest around the exhaust stack and the monitoring point, according to satellite images. The roughness length z_0 was set to 1.5 m.
(2) Displacement height $d_0 = 9$ m.
(3) The plume rise of the emission is calculated according to VDI 3782 Part 3 [65].
(4) Stability class information is derived from wind sigma data of the SODAR measurements at 100 m height. Obukhov-lengths were extracted using the look-up table as described in Ref. [66].

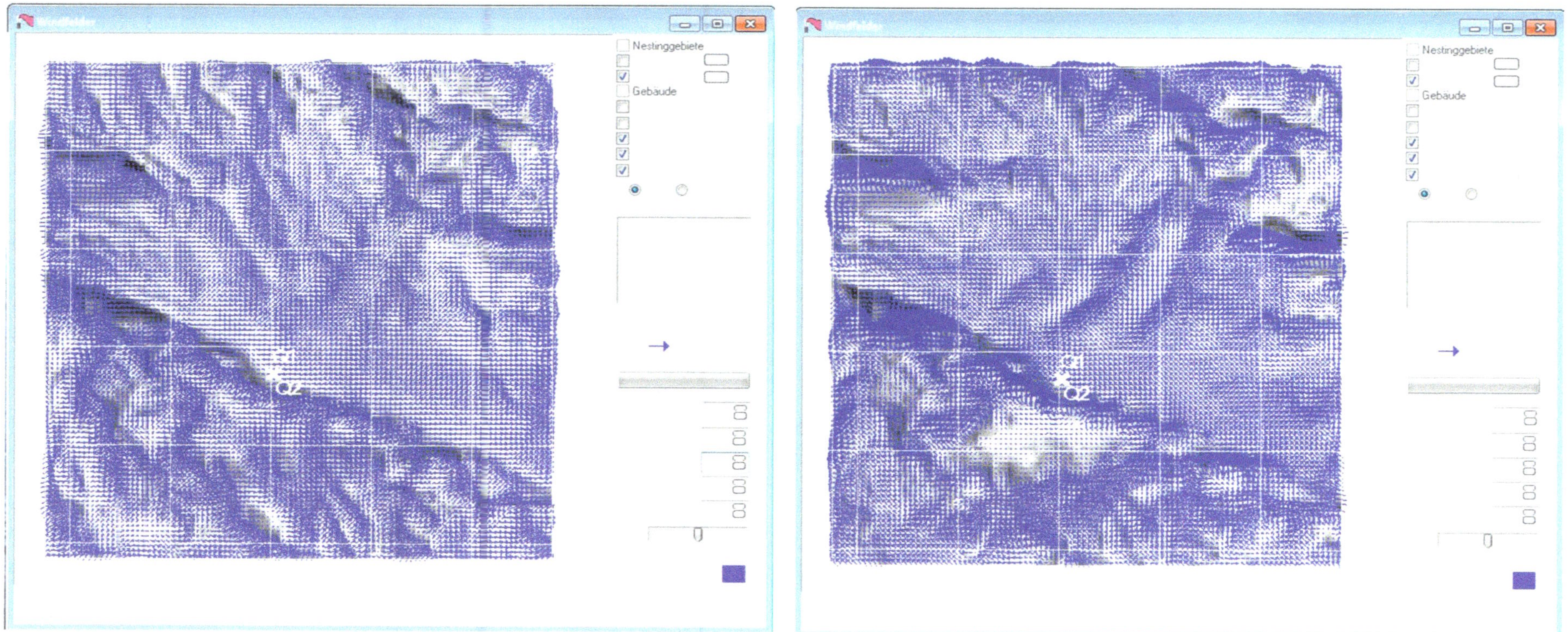

FIG. 123. ARTM model (Bundesamt für Strahlenschutz, Germany) simulation results for the southern (left) and western (right) wind components according to the diagnostic wind field model TalDIA.

A.4.2.3. Input information

Meteorological information was provided on a 30 minute basis at the meteorological station. However, ARTM can deal only with hourly data, and the turbulence parameters are also optimized for hourly data. Temperature and wind data were averaged for an hour (vector averaging for wind). Precipitation data were summed up for each hour.

Differences between the model runs were the usage of the SODAR wind data at 158 m or 376 m altitude above ground. Input information is summarized in Table 27.

A.4.2.4. Results

Figure 124 shows the ARTM simulation results in terms of the predicted ^{35}S activity concentration for the whole period for the assessed wind fields from the SODAR data at both 158 m and 376 m altitude above ground. All other parameters were set the same for both simulations. Weather dependent plume rise was taken into account for both cases.

The time dependent simulations for the validation were conducted considering orography. Figure 124 shows the activity concentration distribution for the same time step of the two simulations at the lowest layer (0–3 m above ground level). In both cases the plume spread to all directions except the sector between east and south. However, the locations of maximum values differ, i.e. in the case of 158 m data it is found around 4000 m away to the west from the source, while for the 376 m data it has a separation of 3500 m in the north direction from the source. In the latter case a larger area is affected with higher concentrations.

A.4.3. Discussion

In the case of the Šoštanj exercise, it was demonstrated that orography has to be taken into account for the simulation. It is an interesting question: whether the simulation using 158 m (approximate stack height) or 376 m (approximate topography height) wind data can provide better agreement with the measurements?

TABLE 27. INPUT INFORMATION USED WITH ARTM IN THE ŠOŠTANJ EXERCISE

Parameter	Windfield estimated from 158 m SODAR data	Windfield estimated from 376 m SODAR data
Horizontal resolution	100 m	100 m
Meteorological data	Transformed to hourly data from SODAR measurements, and precipitation measurements from Šoštanj station	
Plume rise	Yes	Yes
Heat emission (Q)	5.7 MW and 16.2 MW	

Wind field from 158 m data

Wind field from 376 m data

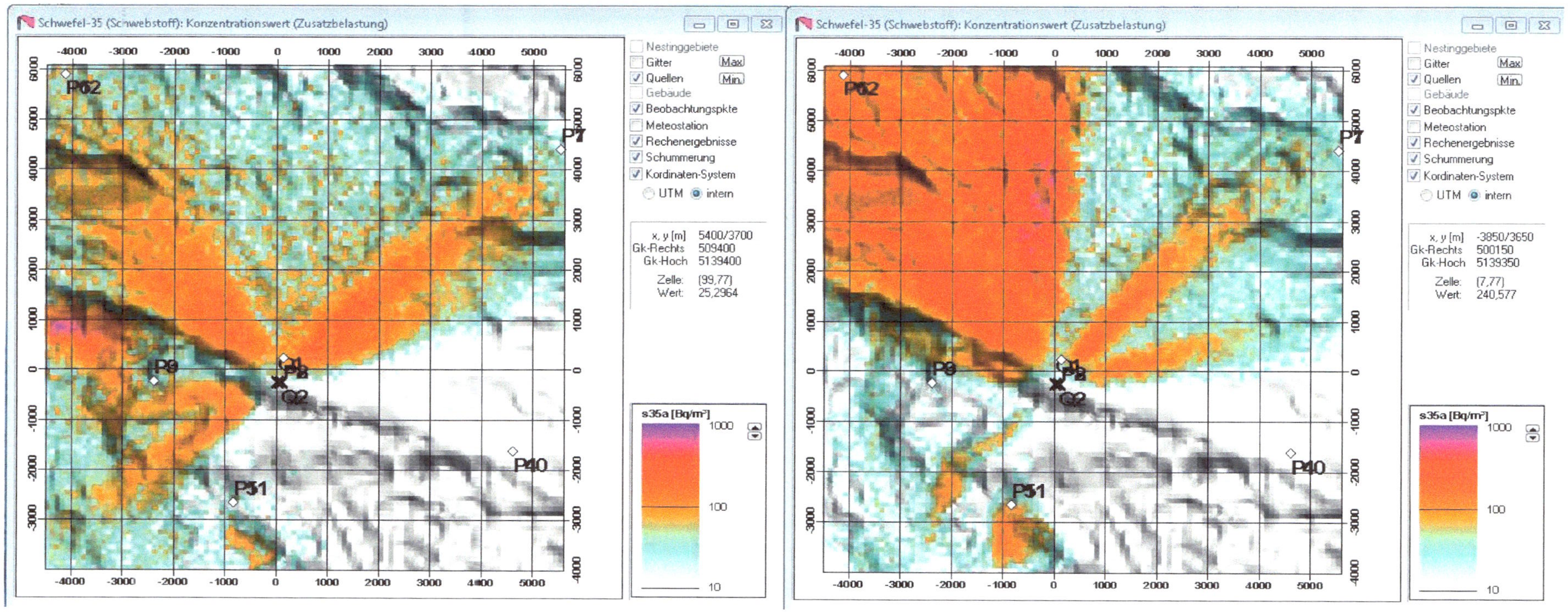

FIG. 124. Simulation results with ARTM model (Bundesamt für Strahlenschutz, Germany) in terms of the ^{35}S isotope. Black Xs denote the locations of the stacks (block 1, 2, 3 and block 4), white diamonds the locations of the monitoring points. Left, wind field based on data from 158 m; Right, wind field based on data from 376 m.

A.5. DESCRIPTION OF ARTM (CHALK RIVER EXERCISE)

The ARTM code was used for the Chalk River modelling exercise by Margit Pattantyús-Ábrahám of the Federal Office for Radiation Protection in Germany.

A.5.1. Introduction

The Atmospheric Radionuclide Transport Model (ARTM) is an atmospheric dispersion model for regulatory purposes. The model is designed for simulating long term atmospheric dispersion of radionuclides from planned releases. The model consists of two subparts:

— Diagnostic wind field model TALdia, providing three dimensional wind and turbulence fields calculated from one measurement point of meteorological data. The turbulence field is optimized for one hour;
— Lagrangian particle tracking according to the results of the TALdia.

During the particle tracking, the radioactive decay, sedimentation, and wet and dry deposition are taken into account if necessary. The model permits the usage of time dependent weather parameters.

At each surface point, the gamma cloud shine is derived from the 3D distribution of the activity concentration, taking attenuation into account. A finite volume source (gamma submersion) term is considered.

The ARTM model does not contain a dose assessment part; however, with its result, the dose received by a reference person can be computed in the case of an annual atmospheric dispersion simulation.

A.5.2. Application to the Chalk River exercise

The orography data around the Chalk River nuclear site was extracted from the Canadian Digital Elevation Model (webservices.maps.canada.ca) as DEM with 20 m resolution. The data were transformed and remapped to the UTM 18North coordinate system. As the first comparison revealed, the inhomogeneity of the terrain had to be considered (see Fig. 125).

A.5.2.1. Key assumptions

(1) The inhomogeneity of the terrain has to be taken into account.
(2) The exhaust stack is at least 600 m away from the buildings of the research site at an elevated point. Therefore buildings were not taken into account for the simulation.
(3) The simulated area was selected in such a way that changes in elevation greater than 1:20 do not occur, because the diagnostic wind field model TALdia cannot cope with greater changes in elevation (i.e. steeper slopes).
(4) The computation of the dose rate at the monitoring point is conducted based on gamma cloud shine results.
(5) The exhaust rate for each week of the experiment is constant.
(6) Dose conversion: 1 Roentgen ≈ 0.01 Sv.

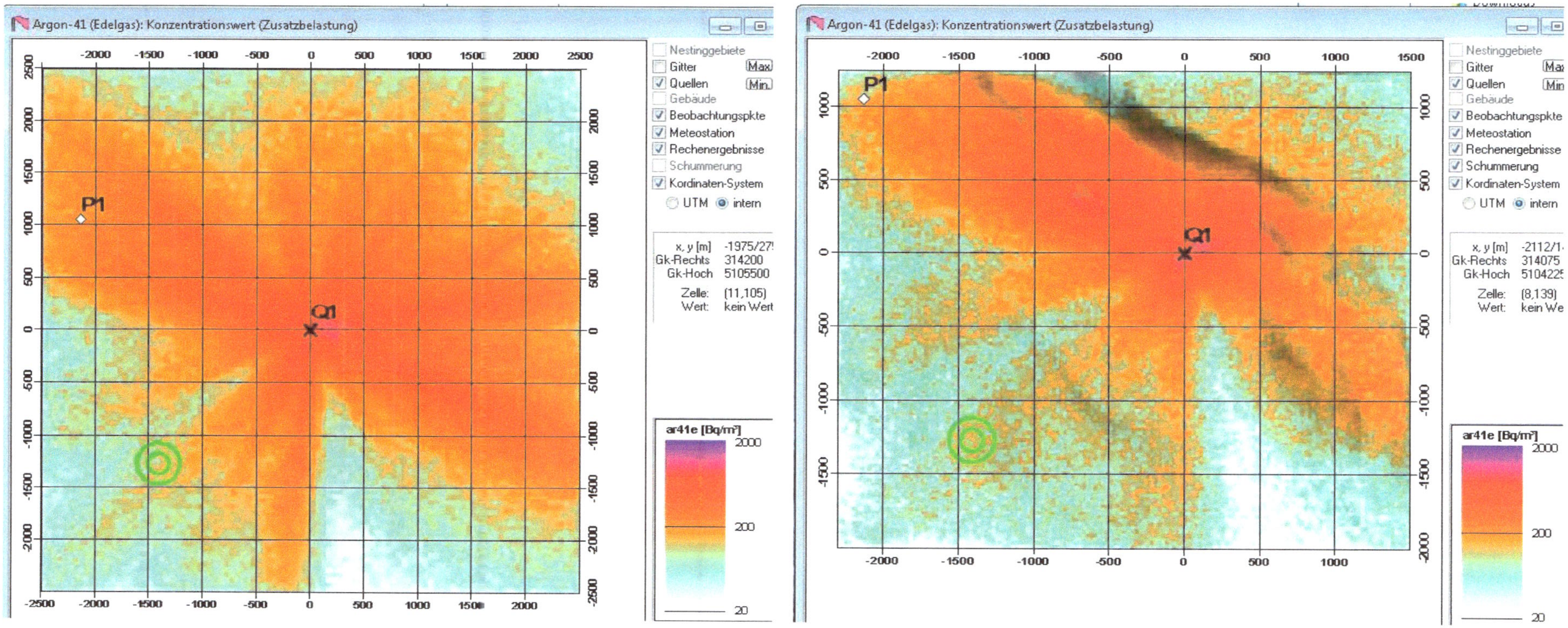

FIG. 125. ARTM model (Bundesamt für Strahlenschutz, Germany) simulation results for ^{41}Ar activity concentrations in Bq/m^3 at the lowest (0–3 m) layer for the whole period. Q1 denotes the source, the centre of green circles the meteorological measurement point, and P1 the location of the monitoring point. Left: simulation considering plain terrain; Right: simulation taking orography into account.

A.5.2.2. *Parameter values*

(1) Land cover is mostly mixed forest around the exhaust stack and the monitoring point, according to satellite images. The roughness length z_0 was set to 1.5 m.
(2) Displacement height $d_0 = 6$ m.
(3) The plume rise of the emission is calculated according to VDI 3782 Part 3 [65].
(4) Dose calculation: Dose = gamma_cloud_shine × time × dose_rate_coefficient (^{41}Ar, Adult) [Sv/h], where time is taken as 3600 s and dose_rate_coefficient (effective_dose, ^{41}Ar, adult) = 4.10×10^{-6} Sv · m^2 · Bq^{-1} · s^{-1}.

A.5.2.3. *Input information*

(1) Meteorological information was provided on a 15 minute basis at the meteorological station. However, ARTM can deal only with hourly data, and the turbulence parameters are also optimized for hourly data. Temperature and wind data were averaged for an hour (vector averaging for wind). Stability classes were determined by the difference between the hourly averaged temperature measurements at 30 m and 2 m above ground. Precipitation data were summed up for each hour.
(2) Differences between the model runs were the usage of time dependent plume rise or no plume rise. Input information is summarized in Table 28.

A.5.2.4. *Results*

Figure 125 shows the ARTM simulation results in terms of the predicted ^{41}Ar activity concentration for the whole period, with and without taking orography into account. All other parameters were set the same for both simulations. Weather dependent plume rise was not taken into account for either of the cases. The effect of the orography is clearly visible on the activity concentration distribution, with the largest differences to the northwest (lower values for plain terrain) and northeast (higher values for plain terrain) of the source.

The time dependent simulations for the comparison were conducted considering orography. Figure 126 shows the activity concentration distribution for the same time step of the two simulations at the lowest layer (0–3 m above ground level). In the case of plume rise, the activity concentration at the lowest level is much smaller, and at the vicinity of the source is practically zero. The emitted ^{41}Ar plume hardly reaches the ground. In the case of no plume rise, the plume reaches the ground rapidly.

Note that the predicted gamma cloud shine at ground level for both simulations shows a very similar pattern.

A.5.3. Discussion

In the case of the Chalk River exercise, it was demonstrated that orography has to be taken into account for the simulation. However, ARTM could not cope with the elevation changes of the original domain; the slopes were too steep for the diagnostic wind field model TALdia. Therefore, part of the original domain was disregarded, in order to be able to run the model.

TABLE 28. INPUT INFORMATION USED WITH ARTM IN THE CHALK RIVER EXERCISE

Parameter	No plume rise	Plume rise
Horizontal resolution	25 m	25 m
Meteorological data	Transformed to hourly data from measurement point	
Plume rise	No	Yes
Heat emission (Q)	0	0.427 MW

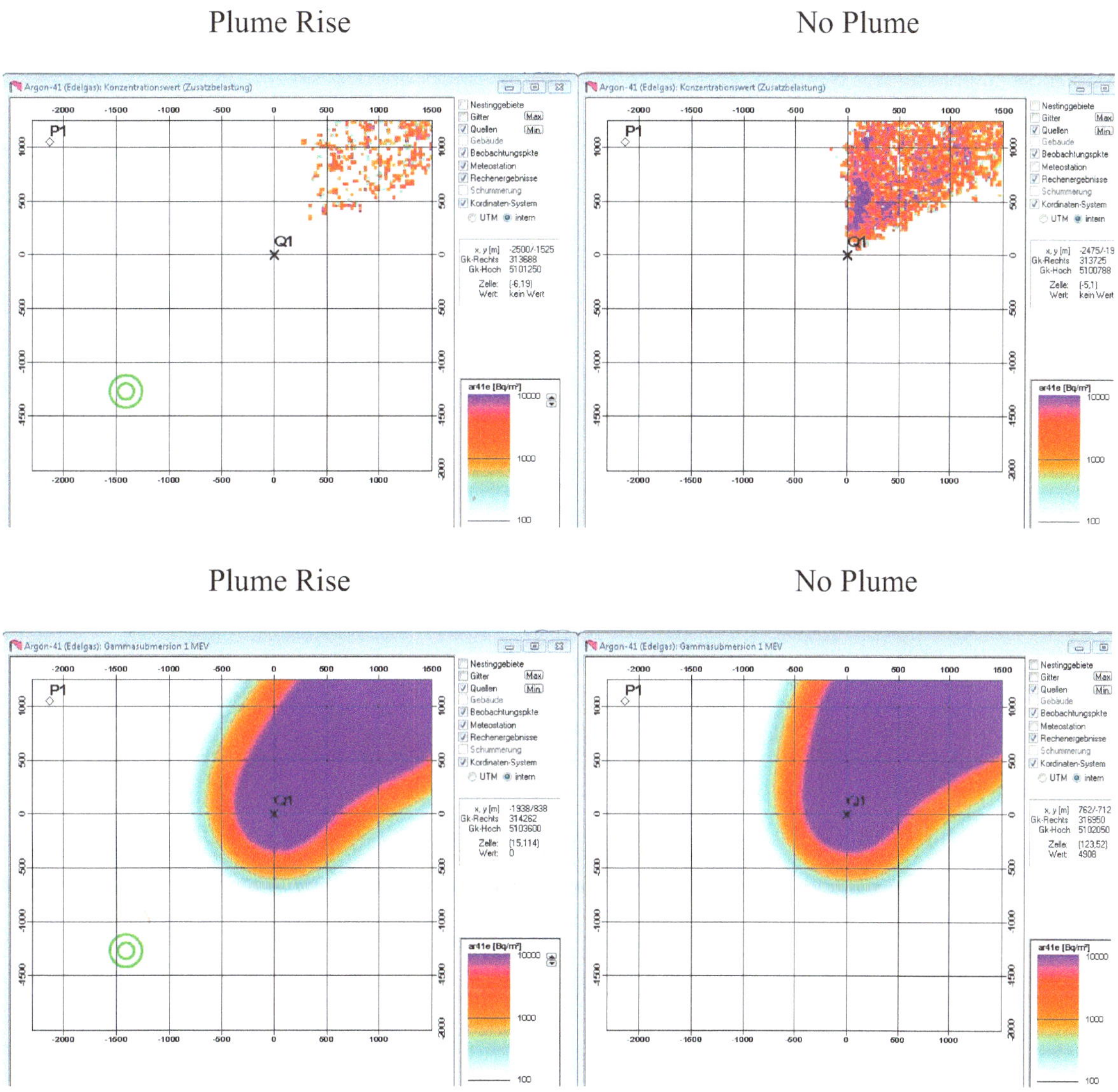

FIG. 126. ARTM model (Bundesamt für Strahlenschutz, Germany) simulation results with and without considering plume rise. Top: ^{41}Ar activity concentration in Bq/m^3 at the lowest (0–3 m) layer; Bottom: ^{41}Ar gamma cloud shine at ground level in Bq/m^3 on 18 July 2008 19:00–20:00 UTC. Symbols are the same as in Fig. 125 (above).

A.6. DESCRIPTION OF JRODOS (CHALK RIVER EXERCISE)

The JRODOS code was used for the Chalk River modelling exercise by Lucia Federspiel from the National Emergency Operations Center (NEOC) in Switzerland.

A.6.1. Introduction

JRODOS is an atmospheric dispersion model for regulatory purposes. It is a decision support system [67, 68] used in Switzerland since 2016. The model is designed for simulating long term atmospheric dispersion of radionuclides from planned releases.

JRODOS contains its own meteorological preprocessor, which prepares the meteorological fields needed by the chosen atmospheric transport model (ATM). There are two ATMs implemented in JRODOS at NEOC for transport modelling on a regional scale (Switzerland and central Europe): LASAT (Lagrangian particle model [69]) and RIMPUFF (puff diffusion model [70]). LASAT is the ATM for operational use at NEOC. The model permits the usage of time dependent weather parameters. MeteoSwiss, the Swiss Meteorological Service, provides the meteorological data (i.e. COSMO 1 for Switzerland and IFS-HRES for central Europe).

At NEOC, transport modelling is optimized for a local and regional scale up to a 3000 km distance from the source. The domain covers the areas where emergency measures might be necessary (Switzerland and neighbouring countries). Transport calculations for distances larger than 3000 km are performed by MeteoSwiss using FLEXPART (Lagrangian particle model) or LAGRANTO (Eulerian particle model).

The aim of participating in this exercise was to determine if the JRODOS version at NEOC can be applied to the Chalk River exercise and to test the predictability and reliability of simulated results compared to the measured data. Since JRODOS at NEOC is mainly optimized for Europe and for simulations with a one week duration, many modifications had to be implemented in order to apply it for the Canadian region and for longer predictions.

A.6.2. Application to the Chalk River exercise

The following subsections describe the applied methodology, in particular, the preparation of the input information and the relevance of these quantities with respect to the final results.

The cloud gamma dose rate and the activity concentration in air (^{41}Ar) at the monitoring point can be directly calculated with JRODOS [71]. The cloud gamma dose rate is given in mSv/h (dose conversion: 1 Roentgen ≈ 0.01 Sv) and the ^{41}Ar activity concentration in Bq/m^3.

A.6.2.1. Source term

The ^{41}Ar source term was provided by the Canadian Nuclear Laboratories (CNL) in Chalk River. A constant exhaust rate for each week of the experiment was assumed.

A.6.2.2. Meteorological data

Meteorological information was provided on a 15 minute basis at the meteorological station. The 15 minute weather data were used to run the first simulation, but some JRODOS limitations were observed. JRODOS cannot deal with too many manual weather inputs (maximum 1000 intervals instead of ~1500 intervals), and two separate simulations had to be performed. Due to the long duration of the simulation (16 days) and the size of the resulting data, the outputs had to be calculated on a one hour time step basis, to avoid further difficulties in the visualization of the results.

For a second simulation, temperature and wind data were averaged for every hour (Fig. 127). Stability classes were determined by the difference between the hourly averaged temperature measurements at 30 m and 2 m above ground (Pasquill-Gifford stability classification). Precipitation data were summed up for each hour.

For a third simulation, MeteoSwiss provided NEOC with the weather data for this time period and region, obtained from the MARS archive (Meteorological Archival and Retrieval System, from the European Centre for Medium-Range Weather Forecasts). The data were provided with a 3 h time resolution (with initialization times at 00, 06, 12 and 18 with +90 h) and with a spatial resolution of only 0.225° (~25 km).

A.6.2.3. Orography

The elevation data around the Chalk River nuclear site were extracted from the SRTM Digital Elevation database by NASA with a 90 m resolution, whereas the land use data were shared with the participants from MEIS (Slovenia) and were derived from the Canadian Digital Elevation Model with a 50 m resolution.

According to satellite images, the land cover is mostly mixed forest around the exhaust stack and the monitoring point. Buildings with height = 15 m and width = 40 m were taken into account but were not expected to influence the simulation due the large distance to the exhaust stack. Indeed, the exhaust stack is at least 600 m away from the buildings of the research site at an elevated point.

All data were transformed and remapped to the UTM 18North coordinate system. The effect of the elevation and land use data on the predicted activity concentrations of ^{41}Ar are compared in Fig. 128.

Figure 128 shows the JRODOS simulation results for activity concentrations of ^{41}Ar in air integrated over the whole period, with and without considering orography. All other parameters were set the same for both simulations. Weather dependent plume rise was taken into account for both cases, with the temperature and wind data averaged hourly.

The effect of the orography is clearly visible on the contour maps of ^{41}Ar activity concentration in air, with the largest differences to the northeast and southwest. For plain terrain, homogeneous and lower values on the north side of the Ottawa River were observed. The orography also determines the lower values observed along the rivers and valleys of the northeast.

From these results, it was evident that the inhomogeneity of the terrain has to be taken into account for the considered simulations.

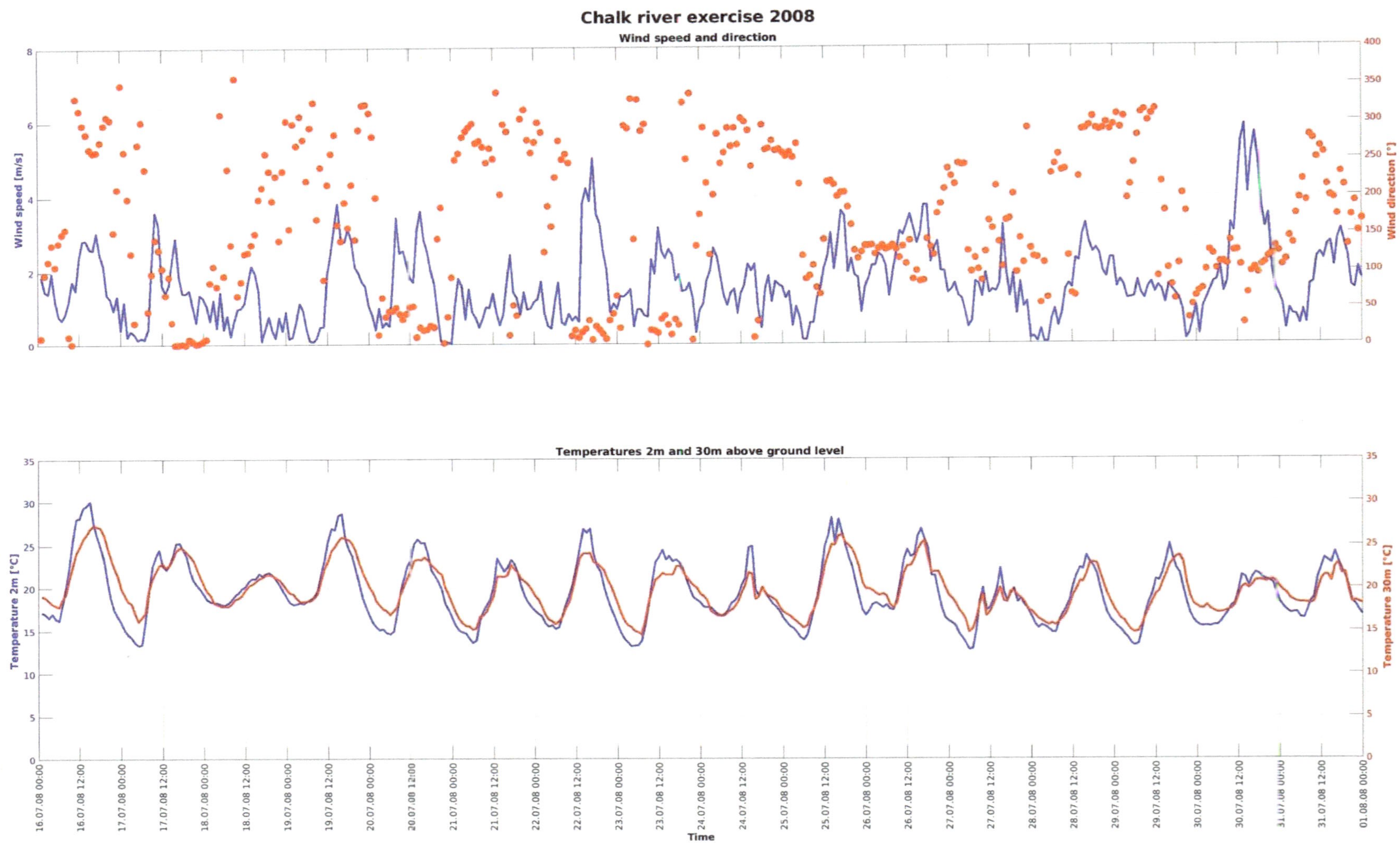

FIG. 127. Top: Measured wind speed (blue, y-axis on left, m/s) and wind direction (red, y-axis on right, °) averaged for every hour. Bottom: Measured temperature at 2 m (blue, y-axis on left, °C) and at 30 m (red, y-axis on right, °C) averaged for every hour.

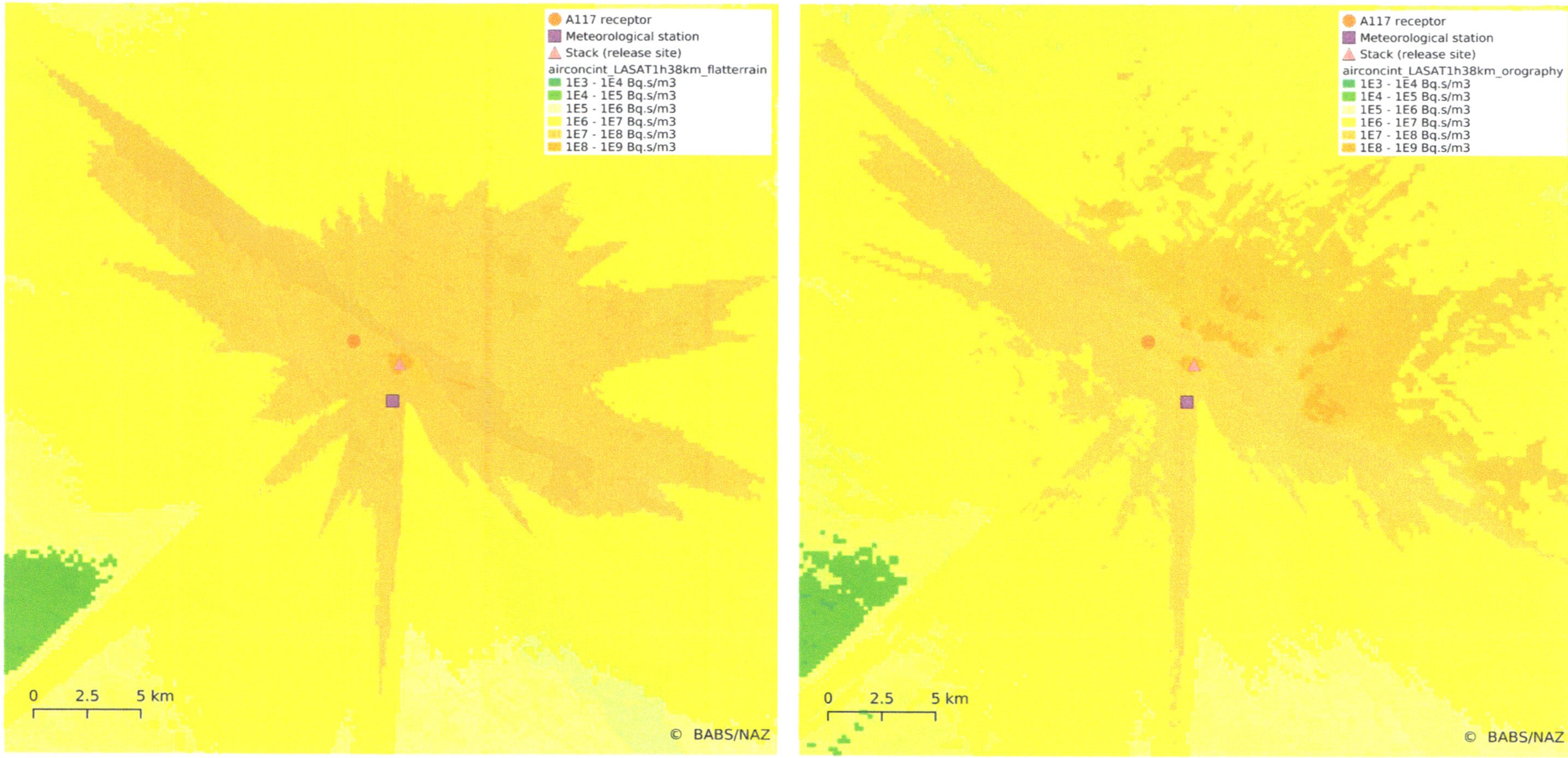

FIG. 128. JRODOS simulation results for activity concentrations of ^{41}Ar in air at 1 m height integrated over the whole period in $Bq \cdot s \cdot m^{-3}$. The pink triangle denotes the source, the violet square the meteorological measurement point, and the orange circle the location of the monitoring point. Left: simulation considering plain terrain, right: simulation taking orography into account.

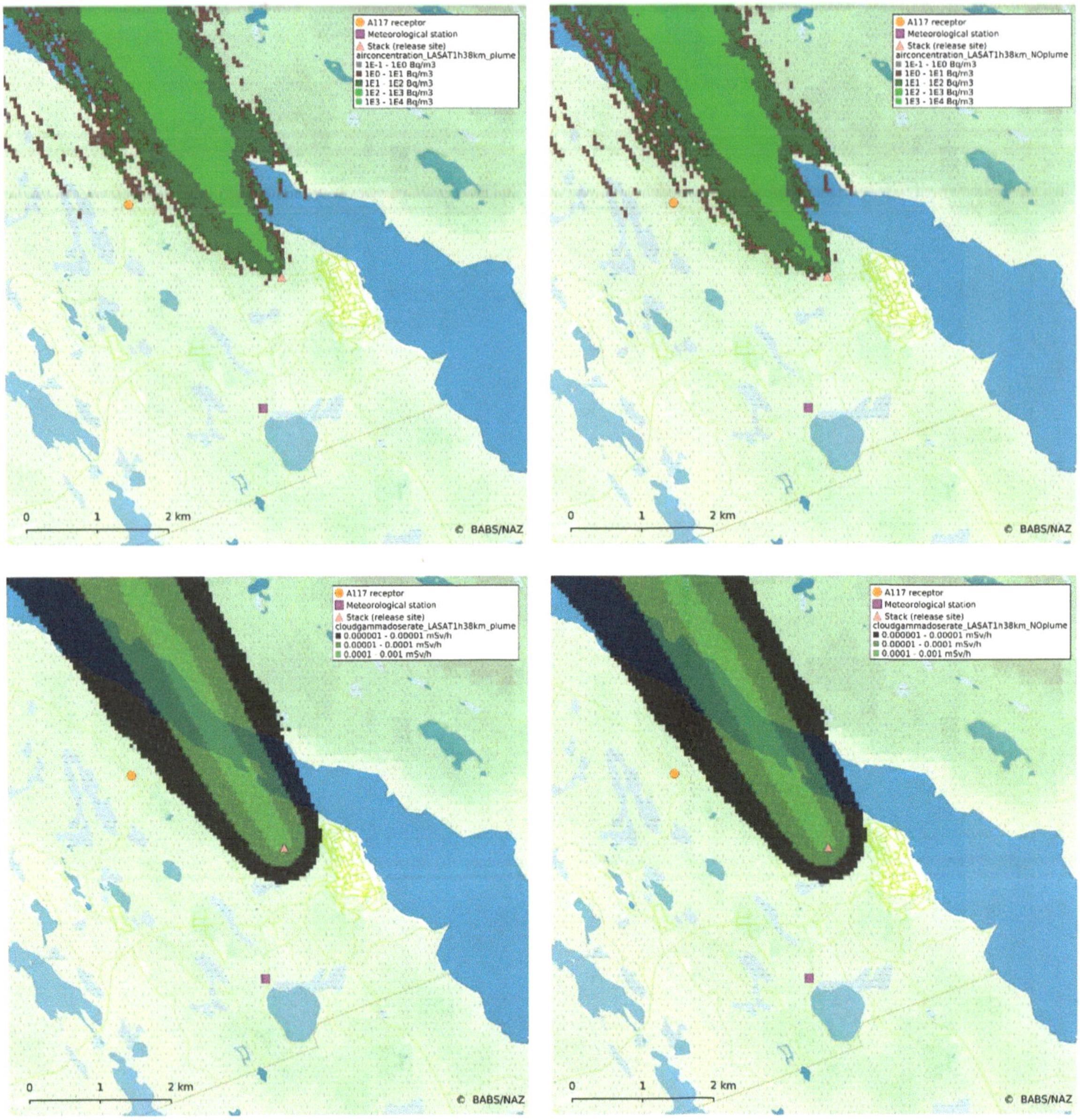

FIG. 129. Simulation results considering plume rise (left) or no plume rise (right). Top: ^{41}Ar activity concentration in air in Bq/m^3 at 1 m height; Bottom: cloud gamma dose rate at 1 m in mSv/h on 18 July 2008 19:00–20:00 UTC. Symbols are the same as in Fig. 128 above.

A.6.2.4. Plume rise

The effect of plume rise was determined during some preliminary simulations (see Fig. 129). For plume rise, the heat emission is calculated according to Ref. [65], with exhaust velocity $v = 10$ m/s and stack diameter $d = 2$ m. The resulting heat emission is $Q = 4.27 \times 10^{+5}$ W.

Two simulation runs were carried out considering orography and land use, with the temperature and wind data averaged for every hour.

Figure 129 shows the contour maps of ^{41}Ar activity concentration in air at 1 m for the same time step of both simulations. In the case of plume rise, the ^{41}Ar activity concentration in air at 1m is smaller, and at the vicinity of the source is practically zero. In the case of no plume rise, the plume reaches the ground more rapidly, and the ^{41}Ar activity concentration values are higher. This difference is less visible for the cloud gamma dose rate at ground level, where both simulations show a very similar distribution.

A.6.3. Discussion

Taking into account the preceding observations, the simulations were carried out considering the plume rise, the elevation and land use data. Table 29 summarizes the five simulations with the corresponding input parameters.

As already explained, the simulations were performed using the weather data on a 15 minute basis, hourly averaged, or taken from the MARS Archive. Four simulations were run with the LASAT model and one with the RIMPUFF model. A grid of 4.8 km radius was considered for one simulation, while for all other simulations a set of four grids with radius up to 38.4 km was taken.

A.7. DESCRIPTION OF LASAIR V. 5 (CHALK RIVER EXERCISE)

The LASAIR v. 5 code was used for the Chalk River modelling exercise by Hartmut Walter and Gerhard Heinrich of the Federal Office for Radiation Protection in Germany.

A.7.1. Introduction

LASAIR v. 5 (Lagrange-Simulation of the dispersion (German: Ausbreitung) and Inhalation of Radionuclides) is a Lagrangian particle model designed for emergency response purposes. The model as used in other modelling exercises has been described previously [2, 11] and elsewhere in this publication.

A.7.2. Application to the Chalk River exercise

Figure 130 shows the geography of the exercise. The symbols and corresponding coordinates (UTM18) for the figure are given in Table 30.

Figure 131 summarizes the land use in the area of the Chalk River exercise. Roughness lengths were set as follows: industrial areas, 1.0 m; vegetated areas, 1.5 m; and water, 0.01m.

Figure 132 shows the buildings in the vicinity of the release point. For this exercise, 197 buildings were selected, with an assumed height of 15 m.

Meteorological data from $t_0 = 0$ to $t_{end} = 15.23:45:00$ were taken from the original data provided to the Working Group in 15 minute time steps. Stability classes were calculated by Margit Pattantyús-Ábrahám. Figure 133 illustrates the anemometer data.

Figure 134 summarizes the data used for the source term and Fig. 135 summarizes other parameters needed by LASAIR for use in the calculations for the Chalk River exercise.

Results of the LASAIR calculation in terms of the gamma radiation dose rate from cloud shine at the receptor location were manually read in 1 hour steps from the LASAIR graphic and compiled in an Excel spreadsheet (Fig. 136). The distance from the source to the receptor was about 2.4 km (Fig. 137).

TABLE 29. SUMMARY OF SIMULATIONS PERFORMED WITH JRODOS

Simulation name	LASAT 1 h 38.4 km	LASAT 1 h 5 km	LASAT 15 min 38.4 km	RIMPUFF 1 h 38.4 km	LASAT MARS 38.4 km
Weather data	Met station, 1 h average	Met station, 1 h average	Met station, 15 min	Met station, 1 h average	ECMWF-MARS archive[a]
Atmospheric dispersion model	LASAT	LASAT	LASAT	RIMPUFF	LASAT
Plume rise	Yes	Yes	Yes	Yes	Yes
Number of grids (rings)	4	1	4	4	4
Spatial resolution, 1° grid	50 m	50 m	50 m	50 m	50 m
Calculation domain	(2 × 38.4 km) × (2 × 38.4 km)	(2 × 4.8 km) × (2 × 4.8 km)	(2 × 38.4 km) × (2 × 38.4 km)	(2 × 38.4 km) × (2 × 38.4 km)	(2 × 38.4 km) × (2 × 38.4 km)

[a] ECMWF-MARS = European Centre for Medium-Range Weather Forecasts-Meteorological Archival and Retrieval System.

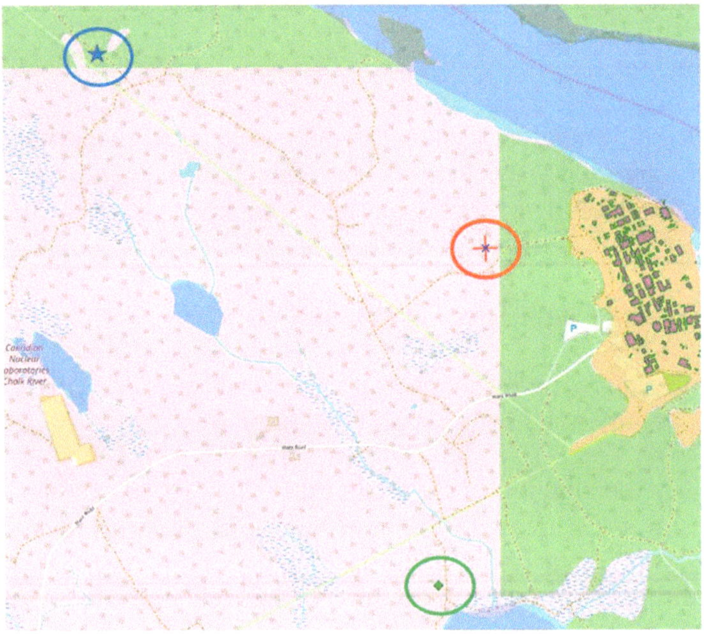

FIG. 130. Locations assumed for LASAIR in the Chalk River exercise. The red cross indicates the stack (release point), the green diamond the meteorological station, and the blue star the receptor. Coordinates for the locations are given in Table 30 (with permission courtesy of Bfs, Germany).

TABLE 30. COORDINATES FOR THE LOCATIONS SHOWN IN FIG. 131

Location (symbol)	x_UTM18	y_UTM18
Stack (red cross)	316197.25	5102770.95
Meteorological station (green diamond)	315869.4	5100973.69
Receptor (blue star)	314059.29	5103838.27

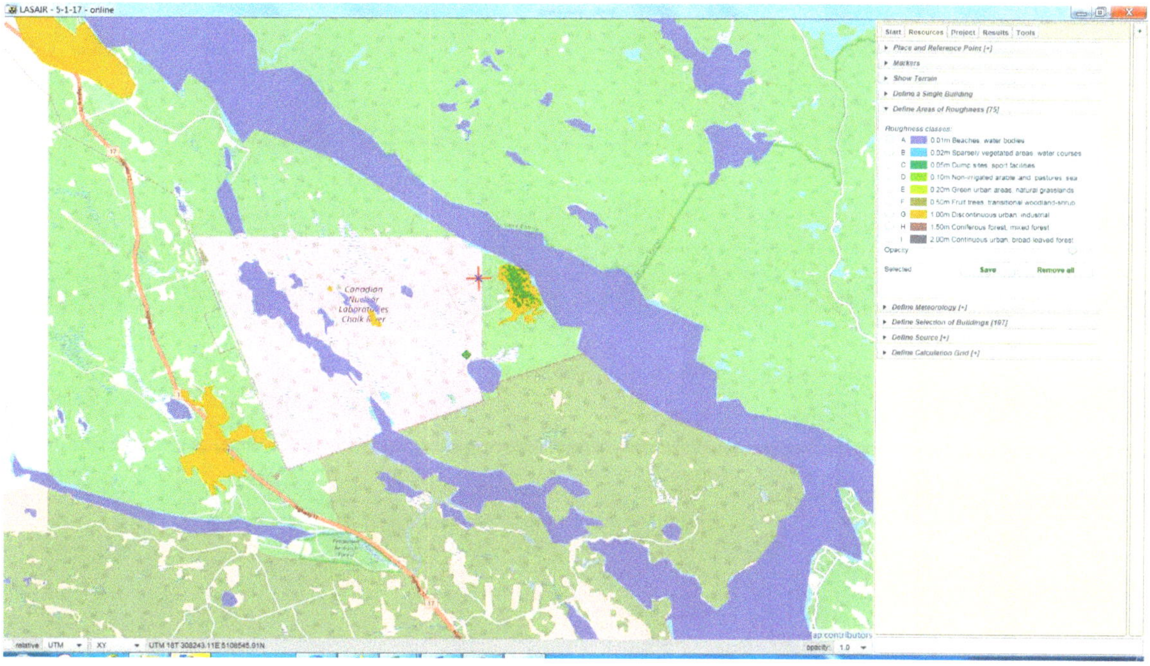

FIG. 131. Schematic map of the land use in the area of the Chalk River exercise. Orange indicates industrial areas, green and grey indicate vegetated areas, and blue indicates water (with permission courtesy of Bfs, Germany).

FIG. 132. Map of buildings in the vicinity of the release point (stack) (with permission courtesy of Bfs, Germany).

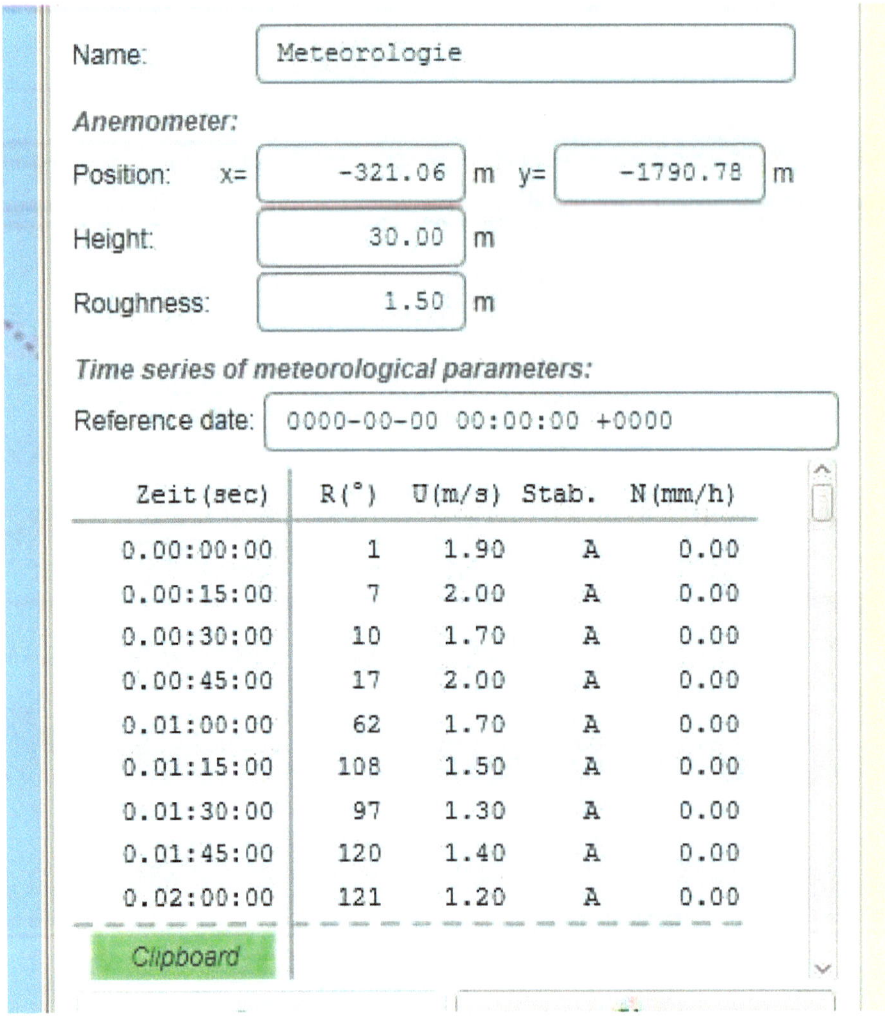

FIG. 133. Example of wind (anemometer) data as used in LASAIR for the Chalk River exercise (with permission courtesy of Bfs, Germany).

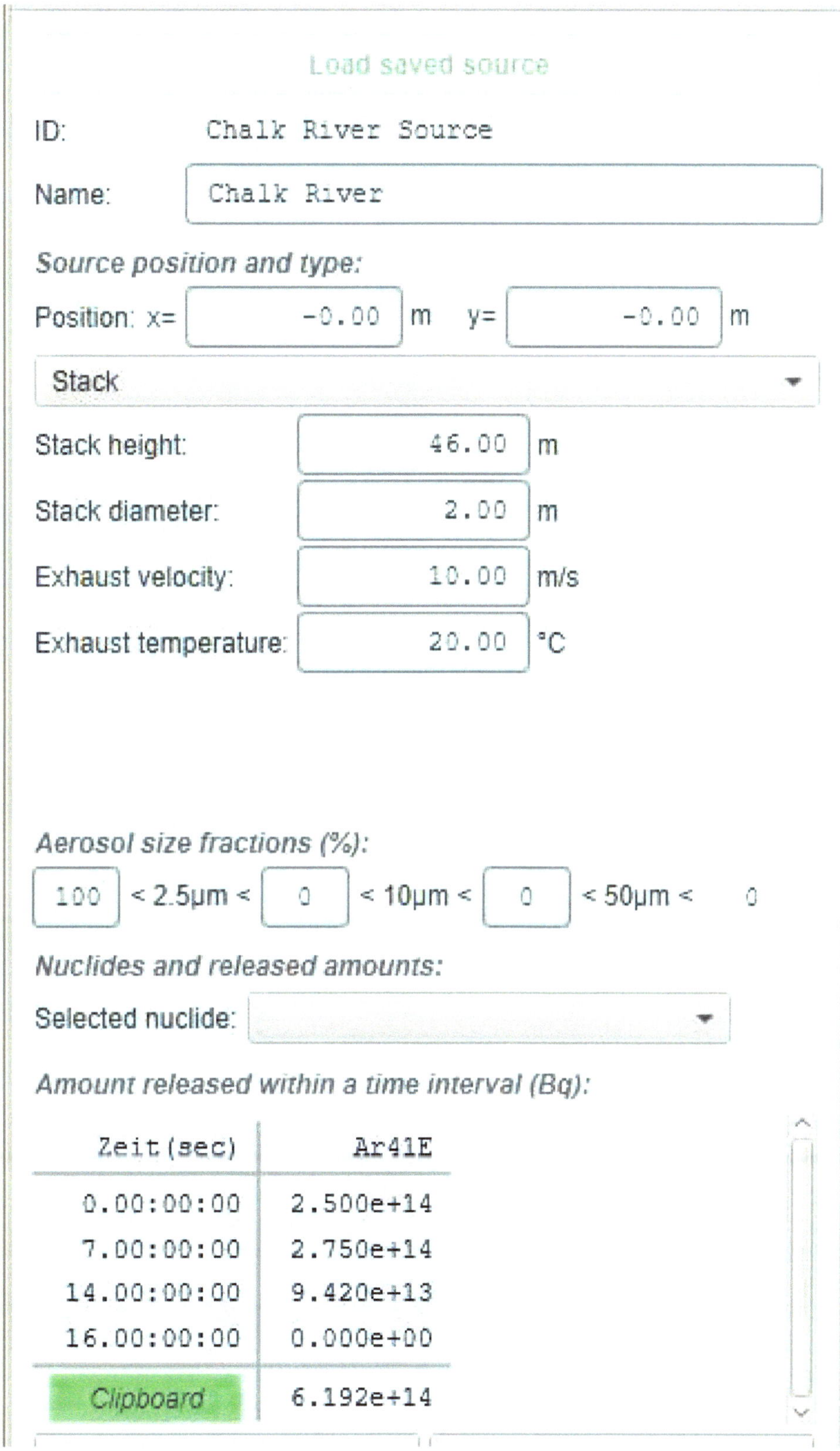

FIG. 134. Summary of source term information used in LASAIR for the Chalk River exercise (with permission courtesy of Bfs, Germany).

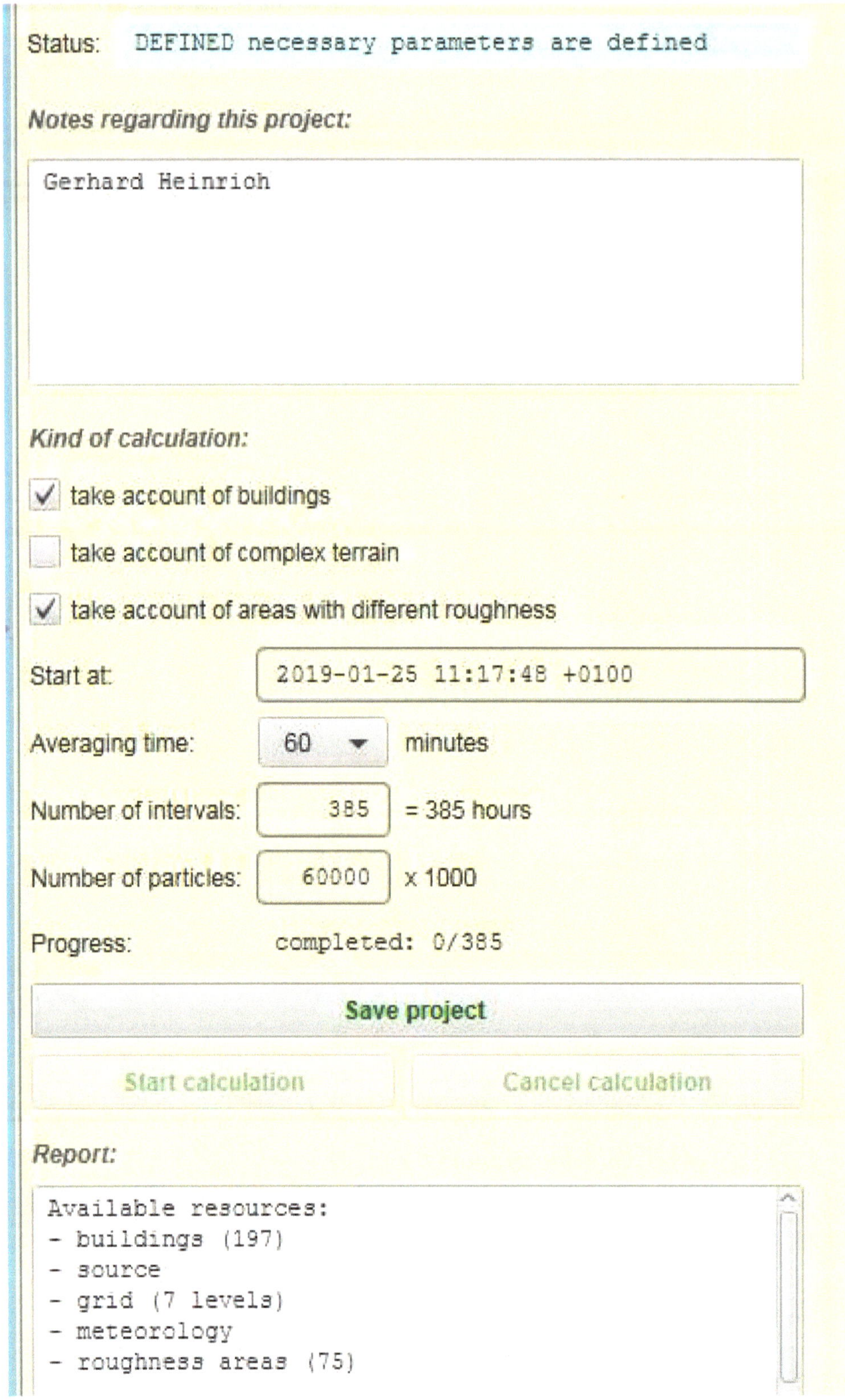

FIG. 135. Summary of input information used in LASAIR for the Chalk River exercise (with permission courtesy of Bfs, Germany).

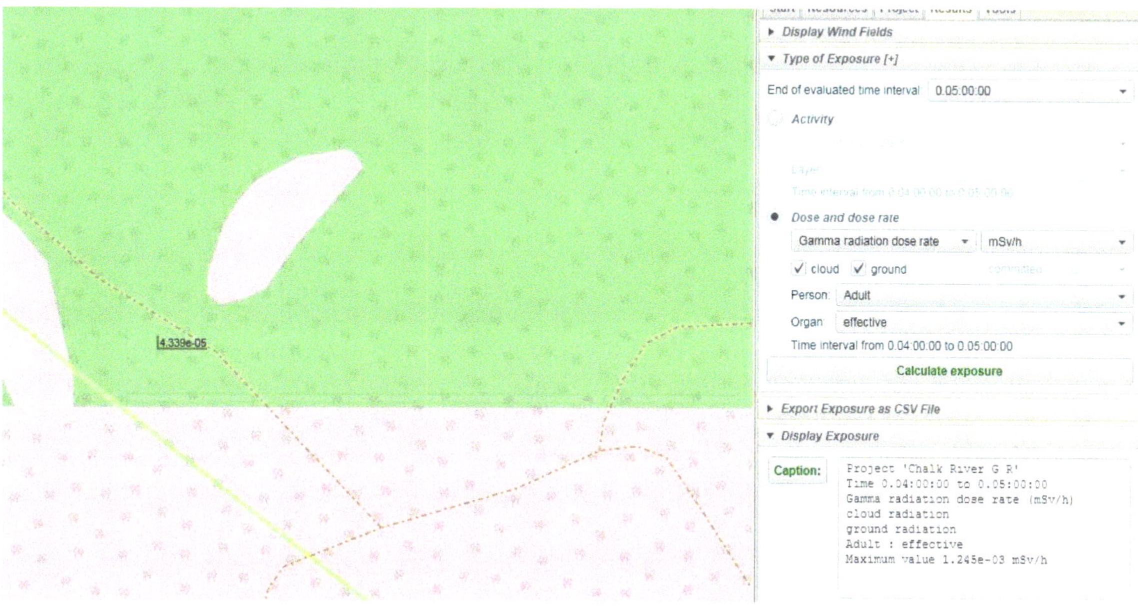

FIG. 136. Example of the graphical results from LASAIR as used for the Chalk River exercise (with permission courtesy of Bfs, Germany).

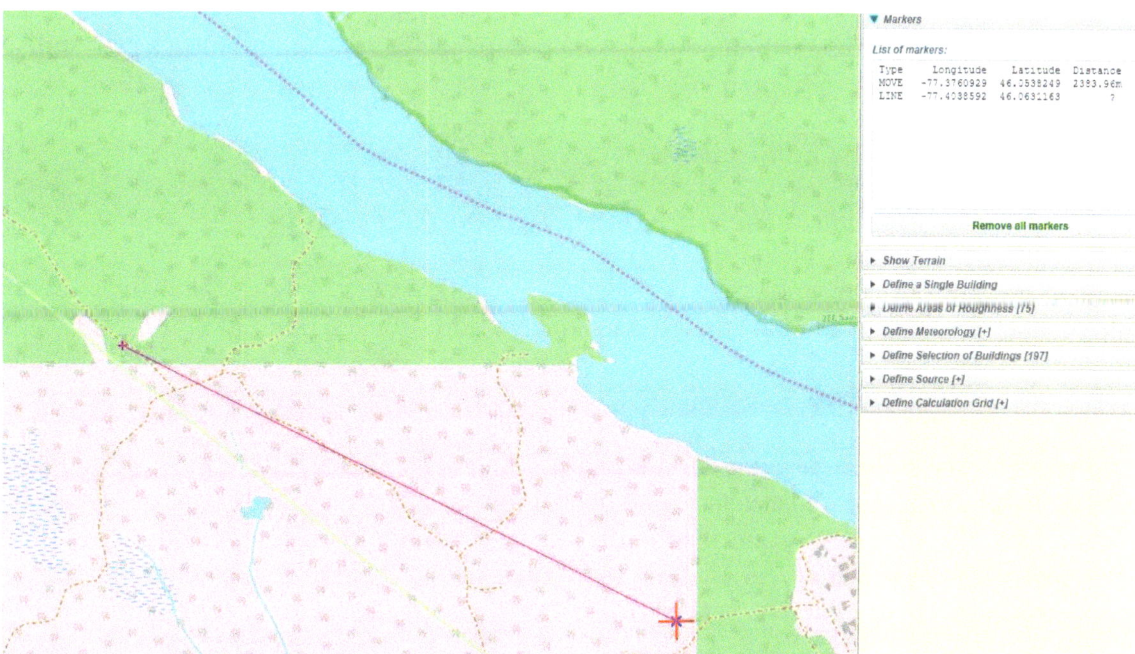

FIG. 137. Estimate of the distance from source to receptor for the Chalk River exercise (with permission courtesy of Bfs, Germany).

A.8. DESCRIPTION OF TINT-UNSCEAR (FUKUSHIMA EXERCISE)

The MS Excel based tool for dose evaluation of a representative person was used for the Fukushima modelling exercise by Kampanart Silva of Thailand Institute of Nuclear Technology in Thailand. In Section 7 of this publication, the MS Excel based tool is referred to as TINT-UNSCEAR.

A.8.1. Introduction

The MS Excel based tool for dose evaluation of a representative person was developed in order to provide a simple tool for assessors with no access to dose evaluation codes so that they are able to evaluate the dose to a representative person (Sv/y) based on the measured or simulated ground concentration of the radionuclides (Bq/m^2). The approach adopted in the model is based on Attachment C-12 to Appendix C of Scientific Annex A of the UNSCEAR 2013 Report [40]; most of the equations refer to Ref. [58].

This is a point estimation tool; thus, it cannot take into account the configuration of the area. The model receives the ground concentration (Bq/m^2) as its input. It is first multiplied by a two exponent function to evaluate the influence of radionuclide migration in soil on the gamma dose rate [40]. The product is then multiplied by the effective dose rate conversion coefficient (nSv/h per kBq/m^2) [71] to obtain the initial effective dose rate. Five one exponent functions are used to represent the attenuation of the activity at the locations in which the representative person stays, including paved surface, unpaved surface, wooden house, wooden fireproof house and concrete building [40]. The user then determines the occupancy factors which indicate the fraction of time that the representative person spends in each location. The sum of the products of each occupancy factor and location factor at each time step is multiplied by the initial effective dose rate to obtain the dose rate at that point in time. Finally, the dose rates are integrated in order to estimate the yearly dose of the representative person. The user can specify the distribution of the ground contamination and occupancy factors and perform a probabilistic calculation using the macro written in VBA.

A.8.2. Application to the Fukushima exercise

A.8.2.1. Key assumptions – characteristics of radionuclides deposited

The composition of the radionuclides deposited follows Ref. [40]. Half-lives of the radionuclides are used to estimate the ground concentration at each point in time.

A.8.2.2. Key assumptions – deposition

As the tool provides only a point estimation, the geometry of the house or the building in Fukushima City was not constructed. The deposition on roofs and walls was not taken into account, as well as the deposition inside the houses or buildings due to the migration of the radionuclides through the air ventilation system. Only the dose from the contamination outside the shelter to the representative person inside the shelter is considered for the indoor dose evaluation.

Although there were ground concentrations for all types of land use and for residential and building areas, only the ground concentrations for residential and building areas are used in this evaluation since the majority of people in Fukushima City live and work in the residential and building areas.

TABLE 31. INITIAL ASSIGNMENT OF ATTENUATION FUNCTIONS FOR EACH LOCATION OF EACH REPRESENTATIVE PERSON

Representative person	Home		Workplace		Others	
	Indoor	Outdoor	Indoor	Outdoor	Indoor	Outdoor
Indoor worker	Wooden house	Paved and unpaved surface	Concrete building	Paved surface	Concrete building	Paved surface
Outdoor worker	Wooden house	Paved and unpaved surface	Wooden house	Unpaved surface	Concrete building	Paved surface

A.8.2.3. Key assumptions – location factors

The locations in which the representative person stays can be categorized as home, workplace and other. Each representative person spends time both indoors and outdoors at each location. The attenuation functions initially assigned to each location for each representative person are shown in Table 31. Both representative persons (indoor worker and outdoor worker) are assumed to live in a wooden house. The ratio between the paved and unpaved surfaces outside the house is set to 59:41 based on the schematic drawing of a typical location in Fukushima City shown in the scenario description. The indoor worker works in a concrete building surrounded by the paved surface, while the outdoor worker works on an unpaved surface and takes rest in a wooden house. As for the remaining time, both representative persons are assumed to spend time in department stores or governmental buildings, which are typically a concrete building surrounded by a paved surface. However, since the attenuation function for paved surface gives a much lower value than that of the unpaved surface, and based on the observations, there is a high possibility that the paved surface around concrete buildings will include unpaved surfaces (e.g. gardens), and so another calculation where all outdoor doses are estimated using the function for unpaved surfaces was performed. The case where all assumptions in Table 31 are applied is hereinafter referred to as Case A, and the case where an unpaved surface attenuation function is used to estimate the outdoor dose is referred to as Case B.

A.8.2.4. Input information – deterministic approach

The calculation is divided into two subcases. In the former subcase (AM subcase), the arithmetic means of the ground concentration for residential and building areas, and of the occupancy factors, were used. In the latter subcase (95th Percentile subcase), the 95th percentile of the ground concentration was computed based on the arithmetic mean and the standard deviation. The 95th percentile value of time spent outdoors and the 5th percentile value of the time spent indoors were adopted for the calculation. When the 5th percentile value was less than zero, it was set to zero. Then the values were normalized in order to make the sum equal to 24 hours.

A.8.2.5. Input information – probabilistic approach

The calculation is divided into two subcases. In the first subcase (normal distribution), normal distributions were applied to the ground concentration for residential and building areas, and for the occupancy factors, using the arithmetic means and standard deviations provided. However, thc distributions were truncated in order not to have any values less than zero. The times were also normalized to make the sum equal to 24 hours. In the latter second (distribution obtained from occupancy survey recommended for the assessment in [14]), the log normal distributions were applied to the occupancy factors of the indoor worker, using the geometric mean and geometric standard deviation, ceteris paribus.

A.8.3. Results

The estimated annual doses for an indoor worker and an outdoor worker for all cases and subcases using the deterministic approach are shown in Table 32. Though the difference in assumption about the location attenuation functions used for outdoor exposure has limited effect on the estimated annual dose of the indoor worker, it has a significant effect on the estimated annual dose of outdoor worker. This could be attributed to the fact that the fraction of the time being outdoors for the indoor worker is much smaller than that for the outdoor worker.

The average and medians of annual doses for the indoor worker and the outdoor worker for all cases and subcases estimated by the probabilistic approach are shown in Table 33. The profile of the annual doses for all cases and subcases of the indoor worker and the outdoor worker estimated by the probabilistic approach are shown in Figs 138 and 139, respectively. It can be observed from Table 33 and Figs 138 and 139 that the difference in distribution type hardly affects the results. The assumption about the location attenuation functions for outdoor exposure has a limited effect on the indoor worker's dose, but a significant effect on the outdoor worker's dose. This is the same as the observation with the deterministic approach. The average (arithmetic mean) and the median (50th percentile) of the indoor worker dose are almost the same as the AM case of the deterministic approach, while they are larger than the deterministic approach for the outdoor dose case. This is due to the truncations of the distribution of the occupancy factors where most of the truncations happened on the lower end of the distribution.

A.8.4. Discussion

A simple tool based on the methodology of UNSCEAR [40] can provide a reasonable estimation of the dose to a representative person using ground concentration as the input when assumptions are appropriate. This approach can be used to estimate the dose during a nuclear emergency even if a calculation code is not available. However, the assessor will need to adequately gather the data and information in order to be able to make rational assumptions.

The assessor performed several cases of parameter survey. In this study, it seems that the type of the distribution (normal distribution versus log-normal distribution) has a limited effect on the estimated annual dose. On the other hand, the truncation of the distribution has a significant impact, especially on the estimated dose to an outdoor worker. The assumption of the outdoor location functions is not so important for the case of an indoor worker's dose, but it led to a notable difference in the case of an outdoor worker's dose.

TABLE 32. ANNUAL DOSES (DETERMINISTIC APPROACH)

Case	Subcase	Indoor worker (mSv/y)	Outdoor worker (mSv/y)
A	AM	0.88	1.11
	95th percentile	1.43	2.27
B	AM	0.90	1.29
	95th percentile	1.56	2.63

TABLE 33. AVERAGES AND MEDIANS OF ANNUAL DOSES (PROBABILISTIC APPROACH)

Case	Subcase	Indoor worker (mSv/y)		Outdoor worker (mSv/y)	
		Average	Median	Average	Median
A	Normal distribution	0.88	0.88	1.20	1.20
	Recommended distribution from occupancy survey [14]	0.87	0.87	1.20	1.19
B	Normal distribution	0.92	0.91	1.39	1.38
	Recommended distribution from occupancy survey [14]	0.91	0.90	1.40	1.40

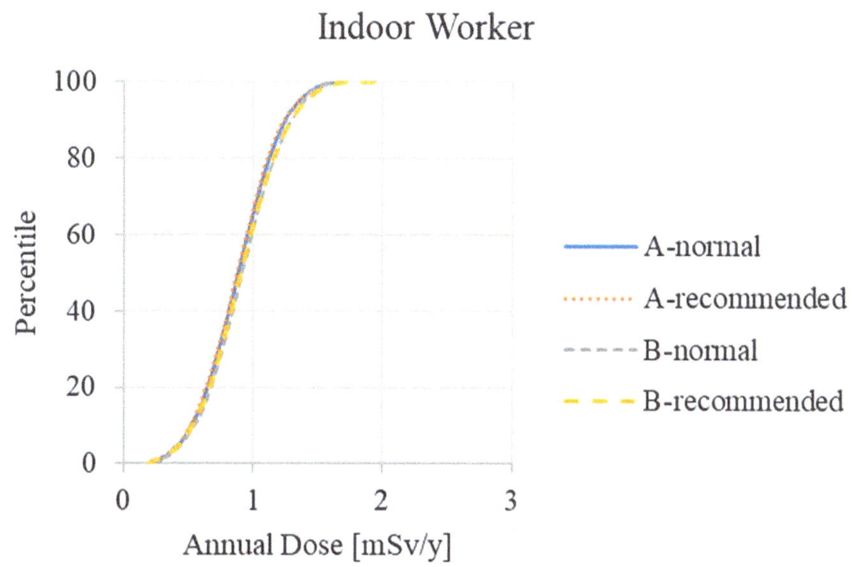

FIG. 138. Probabilistic results for the estimated annual dose to an indoor worker for all cases and subcases, shown in Table 33.

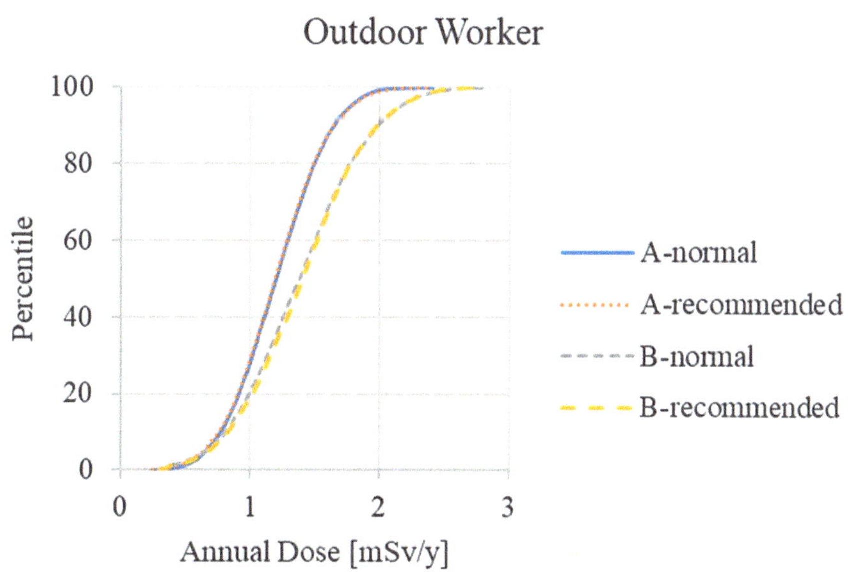

FIG. 139. Probabilistic results for the estimated annual dose to an outdoor worker for all cases and subcases, shown in Table 33.

A.8.5. Discussion

A simple tool based on the methodology of UNSCEAR [40] can provide a reasonable estimation of the dose to a representative person using ground concentration as the input when assumptions are appropriate. This approach can be used to estimate the dose during a nuclear emergency even if a calculation code is not available. However, the assessor will need to adequately gather the data and information in order to be able to make rational assumptions.

The assessor performed several cases of parameter survey. In this study, it seems that the type of the distribution (normal distribution versus log-normal distribution) has a limited effect on the estimated annual dose. On the other hand, the truncation of the distribution has a significant impact, especially on the estimated dose to an outdoor worker. The assumption of the outdoor location functions is not so important for the case of an indoor worker's dose, but it led to a notable difference in the case of an outdoor worker's dose.

A.9. DESCRIPTION OF ERMIN (FUKUSHIMA EXERCISE)

The ERMIN code was used for the Fukushima modelling exercise by Thomas Charnock of the UK Health Security Agency (UKHSA)[8].

A.9.1. Introduction

ERMIN (EuRopean Model of INhabited areas) was developed to predict contamination and residual doses in urban environments and the effect of combinations of cleanup options. The ERMIN model has been developed through a series of EC collaborative projects [44, 52, 72–74].

The main input into ERMIN is deposition onto short grass away from trees, buildings and paved surfaces. The user needs to also provide a description of the ERMIN environment by selecting fractions of different idealized environments from the ERMIN database. Finally, the user can specify the cleanup options that are applied.

ERMIN uses the reference surface deposition and a database of particle and deposition condition dependent empirical ratios to estimate deposition onto other urban surfaces, e.g. paved, roofs, walls, interiors, trees. ERMIN applies empirical functions to calculate the long term retention on these surfaces and uses environment and radionuclide specific dose rate factors to calculate the dose rates from those surfaces to various locations within the environment. Finally, recovery options are represented by removing or moving activity in the environment, modifying retention to account for tie down options, or modifying unit dose rates to account for shielding options.

ERMIN produces several end points. It predicts dose rates and doses in various locations in various environments. By accounting for where people spend time, ERMIN estimates 'normal living' doses with and without recovery options being applied. ERMIN has a database of cleanup options that includes factors describing work rate, cost and waste production. ERMIN can therefore estimate how long an option will take, the doses that workers are predicted to receive, the amount of waste, the waste activity concentration, and the cost.

[8] Formerly Public Health England (PHE).

The current operational version of ERMIN does not account for parameter variability or uncertainty. However, as part of a collaborative EC project CONFIDENCE which focused on uncertainties of emergency management and long term rehabilitation following a nuclear accident, a research version of ERMIN has been developed including distributions for many of the important parameters and the functionality to run ERMIN in a simple Monte Carlo mode to simulate distribution for its main endpoints [73].

A.9.2. Application to the Fukushima exercise

A.9.2.1. Key Assumptions

The key assumptions made to address the Fukushima scenario can be grouped in the following areas:

- The development of a probabilistic model that incorporates the variability of the habits of the population, the variability in the environment, and the regional variation in deposition.
- The development of correction factors to modify the provided aerial survey deposition information of a complex urban environment, to the initial deposition on an idealized grass surface away from buildings, trees, and paved surfaces that ERMIN requires.
- The selection and modification of ERMIN idealized environments to resemble the real Fukushima city environments.

A.9.2.2. Probabilistic model

A simple Monte Carlo approach is used to represent the variabilities that lead to individuals receiving different doses. This approach involves repeatedly sampling the appropriate distributions of parameters and variables, and then running ERMIN to generate large sets of predicted doses from which inferences about the distributions of the real doses can be drawn. For this exercise, the variabilities considered are those associated with:

- The level of contamination an individual is exposed to at home and at work;
- The time spent in different locations (indoors, outdoors, at work and at home);
- ERMIN model parameters, including the initial relative surface deposition and subsequent surface retention parameters.

A.9.2.3. Level of contamination

An individual's dose will depend on the level of contamination within the localities where they spend time. The Fukushima City scenario description provides the results of an aerial survey as a grid of resolution approximately 250 m (see Fig. 140(a), which gives the variation of deposition across the region). It is likely that there is finer scale variation not captured in the survey, but this is probably not important, as people can be expected to move around their local environments and small scale variations will be 'smoothed'. The deposition dataset also gives an indication of the predominant land use in each grid square. Figure 140(b) shows only the grid squares that are classified as predominantly residential, building or road.

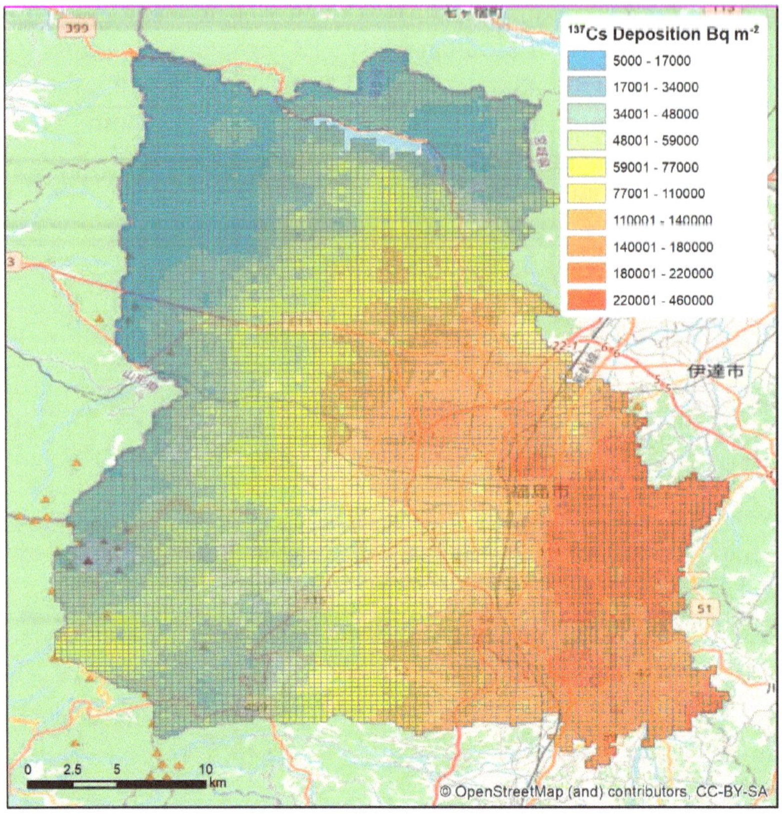

(a)

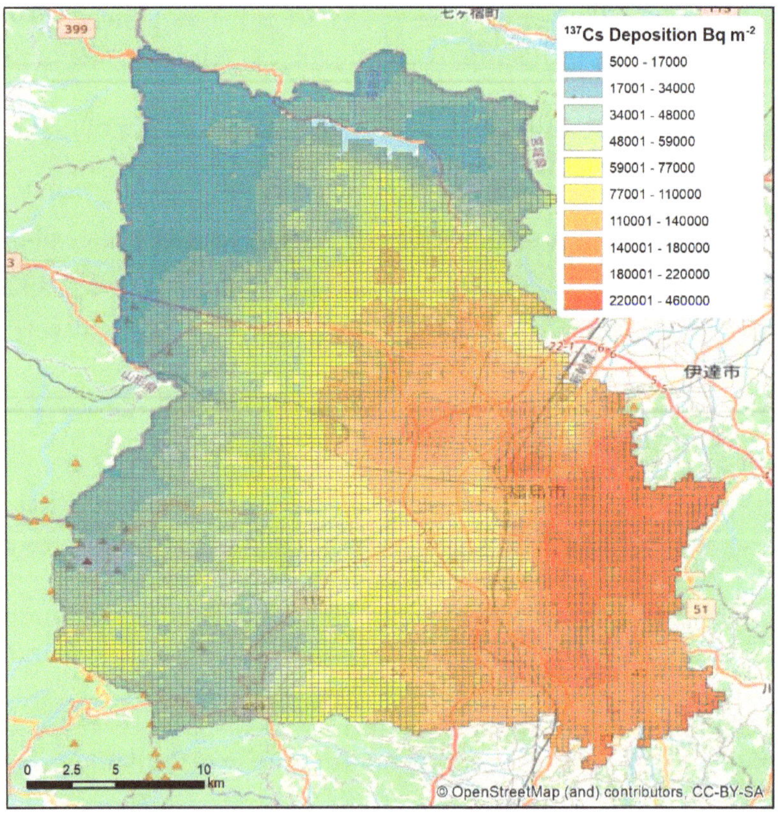

(b)

FIG. 140. Aerial survey data provided in the Fukushima scenario description, converted to a spatial grid and overlaid on OpenStreetMap (data), (a) shows all the survey data and (b) shows just the survey locations over residential areas.

To address the Fukushima city scenario the probabilistic model assumes that an individual spends time at only two locations, i.e. home and workplace. Therefore, each simulation of an individual dose in the Monte Carlo analysis requires two deposition values drawn at random from the set of possible levels (Fig. 141(a)). It would be possible to randomly sample the aerial survey dataset directly, however the residential subset of the data (Fig. 141(b)) is sufficiently normal that it is more convenient to sample this distribution during the Monte Carlo analysis. The sampling is truncated so that values smaller than the minimum value in the dataset or larger than the maximum are discarded.

It is noted that on each iteration, the levels of contamination for the home location and the work location are generated independently of each other; no account is taken of correlations that may result from a tendency for people to live near to where they work. Furthermore, no account is taken of the tendency for urban areas to be zoned into distinct residential, commercial and industrial districts. With this simple model, a person is equally likely to live or work in any grid square.

A.9.2.4. Time spent indoors and outdoors at home and at work

The scenario description provides details of a habit survey on time spent in different locations for two populations, i.e. indoor workers and outdoor workers. The survey has categories of time spent indoors and outdoors in three locations, i.e. home, work and other. For the purposes of the analysis the categories of 'work' and 'other' were conflated into a single 'work' category.

The habit survey also provided summary statistics including the mean and standard deviation of time spent in different locations in hours, and it might be reasonable to assume a normal distribution for these parameters (the data do not exhibit normality, but the sample sizes are small). However, these distributions cannot be sampled naively or independently, as the total hours have to sum to a full day, i.e. 24 hours. To satisfy this constraint, the individual times spent in different locations were converted to fractions, and then normal distributions (truncated between the maximum and minimum for the population) were fitted to these. The Monte Carlo analysis begins by sampling the distribution of fraction of time spent at home, and from this a fraction of time spent at work can be calculated (as 1 – home fraction). Then, independently, the distribution for the fraction of time at home spent indoors is sampled, and from this the fraction of time at home spent outdoors is calculated (as 1 – home indoor fraction). Similarly, the distribution for the fraction of time at work spent indoors was sampled and the time at work spent outdoors is calculated.

By multiplying the fraction of time at home by the fraction of time at home spent indoors, the fraction of the day spent at home indoors is generated, and this can be repeated for fraction of time at home outdoors, and the fractions of time at work indoors and outdoors. This gives four fractions that sum to 1 and divide the day into time spent indoors and outdoors at work and at home. Table 34 shows the statistics for these parameters for the two populations.

A.9.2.5. Uncertainty on deposition and ERMIN retention parameters

Before each run of ERMIN during the Monte Carlo analysis, the distributions of the relative surface deposition and surface retention parameters were sampled independently and provided to ERMIN as input.

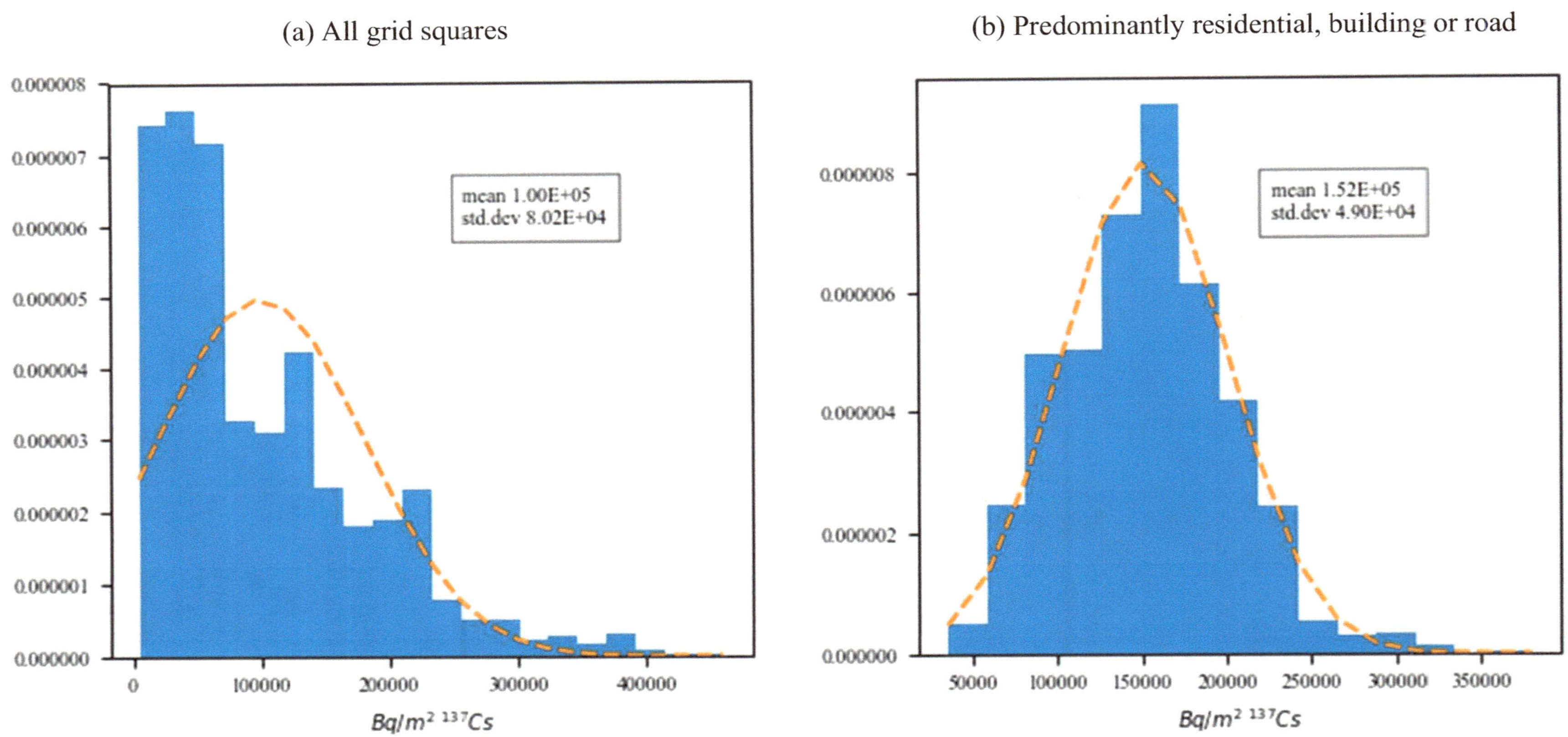

FIG. 141. Normalized histograms of the aerial survey dataset; (a) full dataset and (b) only those grid squares that are predominantly residential, building or road. The dashed line represents a normal distribution with the sample mean and standard deviation.

TABLE 34. STATISTICS REPRESENTING THE DISTRIBUTIONS OF THE FRACTION OF TIME SPENT AT HOME, FRACTION OF THE TIME AT HOME INDOORS, FRACTION OF THE TIME AT WORK INDOORS, CALCULATED FROM THE DATA IN THE HABITS SURVEY IN THE FUKUSHIMA CITY SCENARIO

Worker type and Parameter	Arithmetic mean	Standard deviation	Sample maximum	Sample minimum
Indoor worker:				
Fraction of time spent at home	0.68	0.07	0.76	0.55
Fraction of time at home spent indoors	0.98	0.02	1.0	0.94
Fraction of time at work indoors	0.95	0.04	1	0.88
Outdoor worker:				
Fraction of time spent at home	0.72	0.14	0.97	0.47
Fraction of time at home spent indoors	0.86	0.12	1.0	0.43
Fraction of time at work indoors	0.34	0.24	0.86	0.0

The default ERMIN scenario was assumed; this is suitable for reactor accidents where there are few fuel particles present in the deposition. Under this scenario all the radionuclides except ^{110m}Ag are assumed to be deposited as soluble aerosols. The distributions and parameters compiled by Andersson [75] apply to this physico-chemical form. However, under this scenario (and all scenarios), ^{110m}Ag is assumed to be associated with fuel particles (i.e. in an insoluble form) and consequently during the ERMIN runs default parameters are used to calculate the initial deposition and the retention for this radionuclide which do not account for variability and uncertainty, however ^{110m}Ag contributes very little to the dose and this is not significant.

Relative deposition and subsequent retention parameter distributions are given as means, and standard deviations and are taken from Andersson [75] where available and supplemented with values from the original ERMIN methodology [74]. For the selection of parameters, the following assumptions were made:

- Deposition conditions were assumed to be wet;
- The soil was assumed to be a clay-loam;
- The roof material was assumed to be smooth metal (see below);
- Relative interior deposition was assumed to be described by the mean and standard deviation for a 2.5 m room as calculated in Ref. [73].

The Fukushima City scenario description provides some information on the deposition and retention on both roof and wall surfaces.

The relative deposition of ^{137}Cs on roofs to the initial deposition on the ground is given as 0.02±0.02 in the scenario description. Ref. [75] provides distributions for the parameters that describe the initial distribution and the subsequent retention on several different roof materials (clay tiles, concrete tiles, fibre-cement tiles, silicon coated fibre-cement tiles and metal). Figure 142(a) shows the results of a Monte Carlo analysis in which ERMIN was run repeatedly using parameters drawn from these distributions. In each run the ratio of deposition on the roof at four years to the initial ground deposition (which is taken to be deposition to grass) is calculated, and the combined results for each material are presented as box plots. This exercise shows that the retention on the roofs in Fukushima City most closely resembles the metal material. The scenario description indicates that roofs are made of either metal, slate or mortar (cement). For this analysis metal material was assumed.

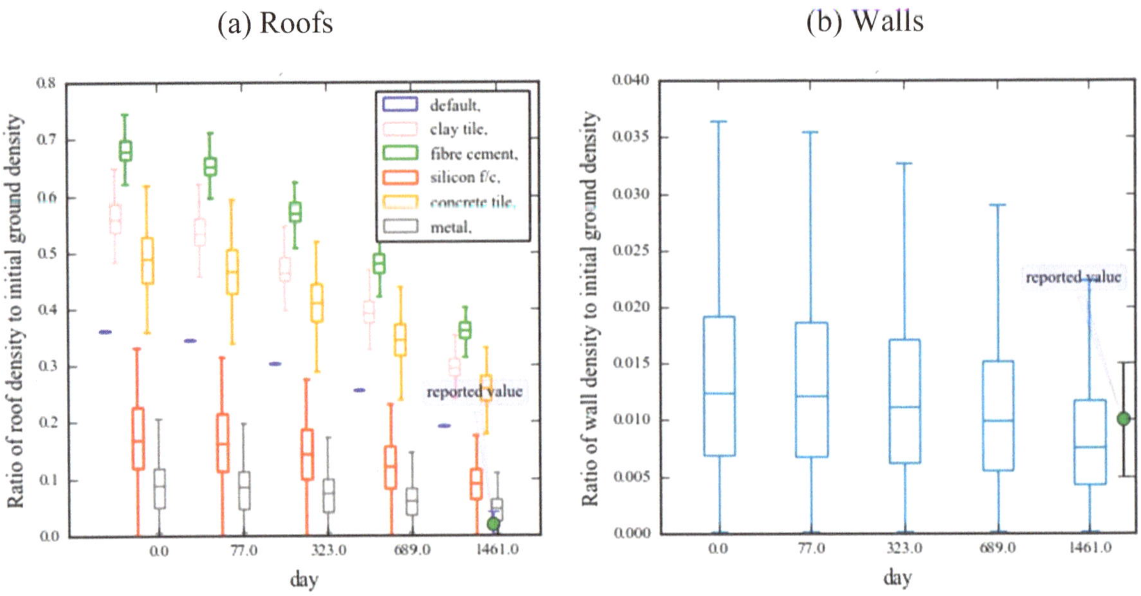

FIG. 142. Box plots showing the result of a Monte Carlo analysis of the predicted ratio of surface deposition on (a) roofs and (b) walls at four years to deposition on ground at time zero. Roof materials are clay tiles, concrete tiles, fibre-cement tiles, silicon coated fibre cement and smooth metal. The results for the current ERMIN default roof are also shown, without uncertainty.

The relative deposition of ^{137}Cs on walls to the initial deposition on the ground is given as 0.01 ± 0.005. Ref. [75] provides parameters for a generic wall, and the results of a Monte Carlo analysis of this are given in Fig. 142(b). The given value matches the predicted values satisfactorily.

A.9.2.6. Deposition to reference surface

The probabilistic approach to treating the provided deposition information is discussed in Section I.9.2.2. This section looks at how the provided deposition data were manipulated so that they could be used within ERMIN.

The principle input into ERMIN is the initial deposition onto a reference surface where the reference surface is a short grass, away from the buildings, paved surface and trees. However, this input is not provided in the Fukushima city scenario description, and furthermore, the study area is built up, with generally small, fragmented areas of grass among a large proportion of buildings and roads. There are very few locations that replicate the idealized ERMIN situation of a large expanse of lawn away from buildings, trees and paved surfaces.

The scenario description provides the results of an aerial survey of ground deposition taken more than a year after the deposition. For each grid square the survey gives a value that reflects the combined deposition of all the urban surfaces, e.g. roofs, walls, paved surfaces, trees, grass, plants and soil, etc., beneath the airborne platform. This is very different from the idealized ERMIN situation, and therefore the aerial survey deposition dataset is not directly usable in ERMIN. This was demonstrated in an initial trial run where the aerial survey data were used as a surrogate for initial deposition onto the reference surface. This gave predictions of public doses that where systematically about a half or third less than those measured.

For the second and final run, correction factors were developed to convert the aerial survey data into something more appropriate for ERMIN. An ERMIN environment is made of different proportions of walls, roof, grass, plants, bare soil, trees, paved surface, and interiors. The simple approach developed for this scenario considers only paved surfaces, roofs and grass surfaces. Trees are omitted because the deposition occurred when there were not many leaves on the trees. Walls and interiors are omitted because deposition onto these is small, particularly under the wet conditions prevailing, and unlikely to register significantly on the airborne instrument. Plants and bare soil surfaces were summed into the total area of grass surface.

As an example of the correction factor calculation, the ERMIN prefabricated/high paved environment is considered. The rough proportions of the grass, paved and roof surfaces for the ERMIN 'prefabricated/high paved' environment are given in Table 35, along with the relative deposition to these surface of ^{137}Cs that ERMIN assumes to apply under wet conditions. Assuming unit deposition (1 Bq/m^2) to the reference surface, the deposition to 1 m^2 of the environment with this mix of surfaces is calculated by multiplying the proportion of surface by the ratio of deposition and summing. In Table 35, the deposition over the complex environment that also includes roofs and paved surfaces is only a fraction of 61% of the deposition on to the reference surface. Assuming the aerial survey picks up deposition on all these surfaces equally and ignoring weathering that may have occurred subsequently, a simple correction factor of 1/0.61 or 1.64 can be derived. Each ERMIN environment used in the analysis will require an individual correction factor because the proportions of urban surfaces are different in each. An environment with a large proportion of grass and few other surfaces will have a correction factor close to 1.0.

A.9.2.7. Environment selection and modification

An important step in an ERMIN analysis is user selection of the environment that most closely resembles the real environment. For the Fukushima City exercise, environments are needed to represent the indoor and outdoor locations of both the indoor worker and the outdoor worker, at work and at home.

The description of a typical residential building in Fukushima City most closely matches the ERMIN prefabricated environment with a high proportion of paved surfaces (shortened to 'prefab' in the following discussion) as both are low shielded buildings. However, there are significant differences, some of which are highlighted in Table 36. For example, the housing density of the Fukushima city is much greater than that assumed in the ERMIN prefab environment.

The environments in the ERMIN library were compiled from previous studies that used Monte Carlo particle transport codes to analyse detailed urban environments [55, 56]. It is not possible to increase the housing density in an ERMIN environment without performing further Monte Carlo analysis, which is beyond the scope of this study. However, it is possible to increase the proportion of paved surface and reduce the proportion of grass surface (as they share a similar geometry), and this was done to produce a modified environment for the Fukushima city scenario (abbreviated to 'prefab.m'); this was used for the home environment indoors and outdoors for both the indoor worker and outdoor worker.

TABLE 35. EXAMPLE CALCULATION OF A SIMPLE CORRECTION FACTOR FOR THE 'REFABRICATED/HIGH PAVED' ENVIRONMENT

Surface	Proportions of surface in the prefabricated environment	Ratio of initial deposition on surface to deposition reference surface	Total deposition to the surface in 1 m^2 of mixed prefabricated environment, assuming 1 Bq/m^2 to the reference surface (Bq)
Roof	0.12	0.085	0.001
Total paved (sum of roads, pavement and other paved)	0.29	0.01	0.0029
Total grass (sum of grass, plant and bare soil)	0.59	1	0.59
Total prefabricated environment deposition			0.61 Bq/m^2

TABLE 36. DIFFERENCES BETWEEN THE TYPICAL FUKUSHIMA RESIDENTIAL ENVIRONMENT AND THE ERMIN PREFABRICATED/HIGH PAVED (PREFAB) ENVIRONMENT

Parameter	Typical Fukushima city residence	ERMIN prefab environment
Distances between houses	5 m, 5 m, 10 m,	~10m, ~30m
Total buildings area	40%	>10%
Paved area/total outside area	Mostly paved	Mostly non-paved
Floors	2	1

The scenario description does not give a typical workplace environment, but it is reasonable to assume that for the indoor worker, this is likely to be a more robust structure than the residential building. Therefore, for the indoor worker the ERMIN 'multi-storey building/high paved' was chosen (abbreviated to 'multi') and used both for the indoor environment and the small amount of time they spend outdoors at work. For the outdoor worker, the choice of indoor work building is much less important as so little time is spent there; therefore the modified prefabricated environment (prefab.m) was selected. For outdoor locations at work, the 'open area/park' ERMIN environment was chosen (abbreviated to 'open'); this is an environment with no buildings, but with some trees and paved surfaces.

The environments chosen to represent the time spent indoors and outdoors at home and at work are summarized in Table 37.

As noted above, each environment requires a correction factor to adjust the amount of deposition on the reference surface (grass) so that the total amount of deposition in the mixed environment more closely matches the aerial survey. The ERMIN environments and their correction factors are given in Table 38.

A.9.3. Results

Table 39 contains the ERMIN predictions for the two populations. These have been extracted from a Monte Carlo analysis. Figure 143 shows histograms of the dose to the two populations from the same analysis. As expected the indoor workers tend to receive lower doses than the outdoor workers, but there is overlap between the two populations. The outdoor workers have a wider distribution, probably due to there being a wider variation of occupancy; compare for example the maximum and minimum fractions of time spent indoors at home between the two populations in Table 34 (the indoor worker varies between 1 and 0.94, and the outdoor worker varies between 1 and 0.43).

TABLE 37. A SUMMARY OF THE ERMIN ENVIRONMENTS USED TO REPRESENT INDOOR AND OUTDOOR LOCATIONS, AT WORK AND AT HOME, FOR THE TWO WORKER TYPES (INDOOR WORKER AND OUTDOOR WORKER) IDENTIFIED IN THE FUKUSHIMA CITY SCENARIO

Worker type	Home indoor environment	Home outdoor environment	Work indoor environment	Work outdoor environment
Indoor worker	Modified prefabricated environment (prefab.m)	Prefab.m	Multi-storey building/ high paved (Multi)	Multi
Outdoor worker	Prefab.m	Prefab.m	Prefab.m	Open area/park (Open)

TABLE 38. ERMIN ENVIRONMENTS AND CORRECTION FACTORS USED IN THE FUKUSHIMA CITY EXERCISE

ERMIN environment	Label	Correction factor
Prefabricated environment with a high proportion of paved surface and no trees	Prefab	1.64
Modified prefabricated environment with 40% paved surface and no trees	Prefab.m	3.01
Multi-storey environment with a high proportion of paved surface and no trees	Multi	3.04
Open area, park like environment with a small proportion of trees and paved surfaces and no buildings	Open	1.11

TABLE 39. PREDICTED DOSE FOR THE PERIOD FEBRUARY 2012 TO JANUARY 2013

Worker type	Percentiles (mSv)				Arithmetic mean (mSv)	Standard deviation (mSv)
	5th	50th	75th	95th		
Indoor worker	0.44	0.90	1.1	1.6	0.94	0.34
Outdoor worker	0.95	1.8	2.3	3.1	1.9	0.65

A.9.4. Discussion

Figure 144 shows the predicted contribution to annual dose from different surfaces for the 'prefab.m' environment. As discussed in Section I.9.2.7, this environment was modified to increase the amount of paved surfaces to more closely match the description of the residential environment in Fukushima City; however, the density of buildings is less than the typical residential environment described in the Fukushima City scenario description. In the period of interest (the third column in the graph) approximately 30% of the dose is from paved surfaces and about 10% from roofs, with the remainder largely coming from soil surfaces (grass, plants and bare soil) and the other surfaces negligible. In the real residential environment of Fukushima city, it would be expected that more would come from exterior building surfaces (roofs and walls) and less from the soil surfaces.

A.9.5. Acknowledgements

ERMIN was developed under a series of EC funded projects.

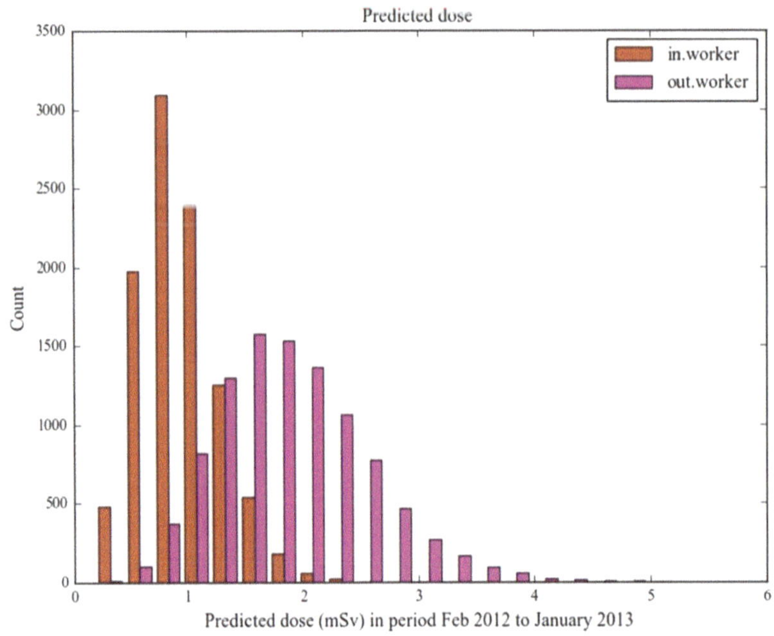

FIG. 143. Histograms of the ERMIN predicted dose for indoor workers and outdoor workers for the period Feb 2012 to January 2013.

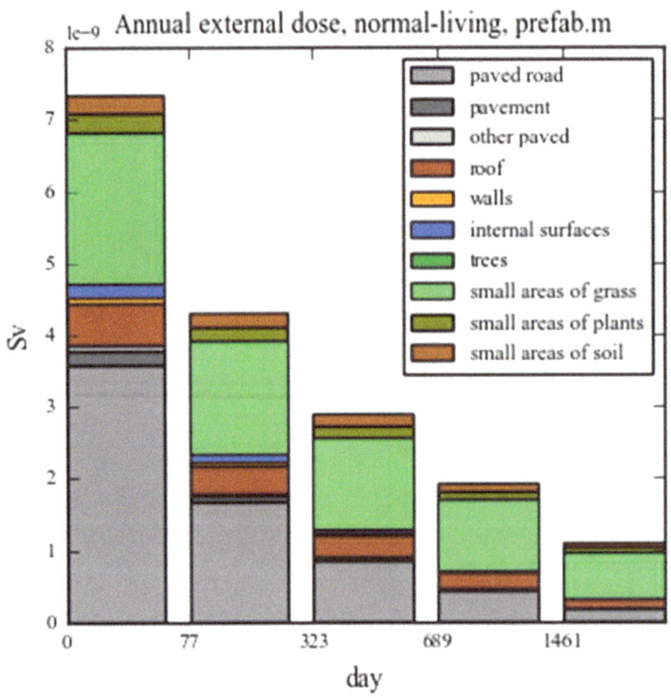

FIG. 144. Predicted contribution by surface to annual dose for an individual living permanently in the prefab.m environment and spending 98% of their time indoors. The doses have been normalized to 1 Bq/m² deposition on the reference surface and need to be interpreted as being integrated from the time on the x-axis for one year, i.e. the first bar represents the predicted annual dose from time 0 to 1 year. The third bar represents the predicted annual dose from day 323 to 688 and corresponds to the period of the exercise endpoints.

A.10. DESCRIPTION OF METRO-K (FUKUSHIMA EXERCISE)

The METRO-K code was used for modelling of the population doses in an urban area based on the Fukushima experience by Joeun Lee of Korea Atomic Energy Research Institute in the Republic of Korea.

A.10.1. Introduction

METRO-K (Model for Evaluating the Transient Behavior of RadiOactive Materials in the Korean Urban Environment) is a dynamic model which has been developed for the prediction of the total exposure dose with time following a contamination event. Three radionuclides (Cs, Ru, and I) and three types of iodine (elemental, organic, and particulate forms) are considered in METRO-K. Many of the parameters used in METRO-K depend on the radionuclides' properties and on the types of surface on which radionuclides are deposited. As shown in Fig. 145, METRO-K predicts the exposure dose in an urban environment starting with the radionuclide activity concentration in air. The surface contamination through dry and wet deposition can be estimated with empirical parameters. The radionuclide activity concentrations on the surfaces are modified by considering weathering processes such as wind, pedestrians, traffic and migration into soil. The absorbed dose is calculated using predetermined values of 'air kerma', which were originally derived by Mechbach et al. [56] and modified for application to the Korean environment [76]. Finally, the total exposure dose for a specified location is evaluated by summing the exposure doses resulting from each contaminated surface.

A.10.2. Application to the Fukushima Experience Modelling Exercise

A.10.2.1. Key assumptions

The data given in the exercise are the initial caesium deposition (activity concentration) in May 2012 and the deposition ratio of each radionuclide normalized by the caesium activity concentration on 1 January 2013. In order to determine the initial concentration of total radioactivity, which is the starting point of the scenario assessment, the initial activity concentration of caesium was deduced backwards from 1 January 2013, and the initial deposition (activity concentration) of each radionuclide was obtained by multiplying by the normalized ratio. After the initial deposition, the deposited radioactive material and its radioactivity are reduced by radioactive decay and weathering; the weathering process after initial deposition is represented in Eq. (3) below. Based on the above procedures, the surface specific radionuclide activity concentrations of the houses for the particular period were obtained:

$$A(t) = A_0(t)\left(a_1 e^{-ln2\frac{t}{\tau_1}} + a_2 e^{-ln2\frac{t}{\tau_2}}\right), \text{with } (a_1+a_2 = 1) \quad (3)$$

where:

$A_0(t)$ is the radionuclide activity concentration at time t including decay but without weathering;
$A(t)$ is the radionuclide activity concentration at time t with the effects of weathering applied;
a_1 and a_2 are, respectively, the short term and long term constants (the applicable fractions for short term and long term weathering);
τ_1 and τ_2 are the short term and long term weathering half-lives.

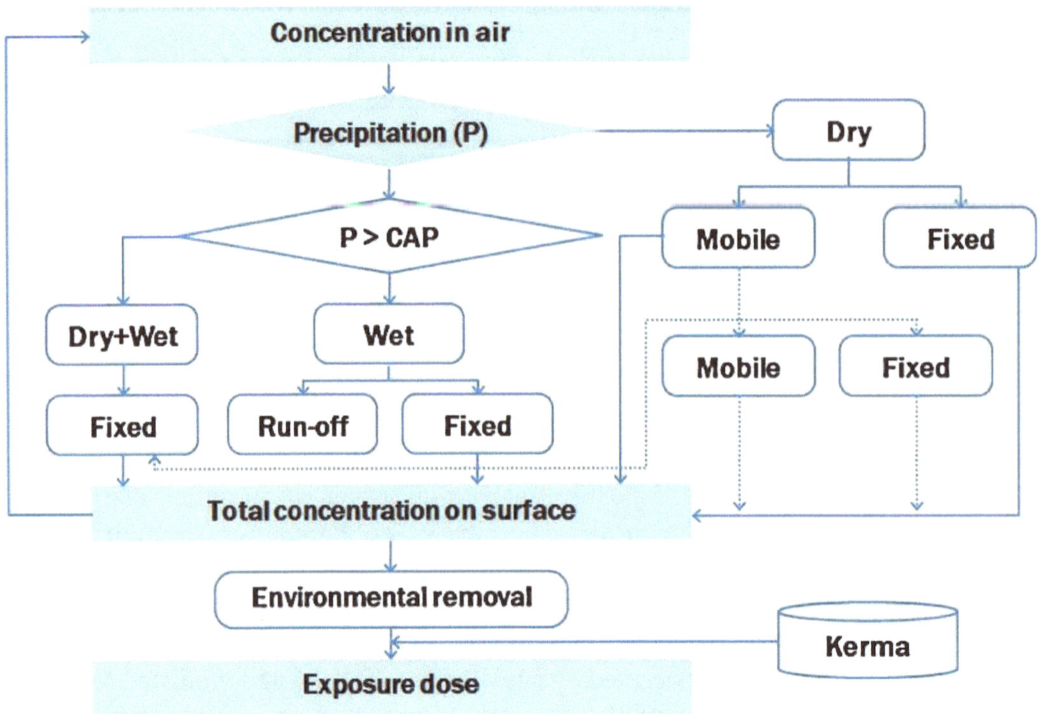

FIG. 145. Schematic diagram of METRO-K. (CAP is the Critical Amount of Precipitation and represents the minimum precipitation amount that occurs during a runoff.)

In addition, after the activity concentrations of radionuclides on each surface are obtained, the external dose from each surface is evaluated by multiplying by the 'kerma' value. METRO-K includes the kerma value, which is calculated by the Monte Carlo method considering properties of typical Korean residential houses. However, since the Korean residential houses are usually made of reinforced concrete, the kerma values in the latest version of METRO-K do not match with the exercise. In order to consider the building characteristics of wooden houses in Fukushima city, previous versions of kerma values were used for the first floor house that were originally derived by German researchers.

A.10.2.2. Parameter values

By applying the assumptions mentioned in Section I.10.2.1, the radionuclide activity concentrations of the initial deposition in March 2011 were obtained for each radionuclide, before weathering, and the initial deposition of each radionuclide for each surface type is shown in Table 40. After the initial deposition, the decrease of the radionuclides' activity concentrations due to the weathering effect follows Eq. (3), where A_0 is the radioactivity decaying over time without weathering. Due to the relatively short half-life of iodine isotopes compared to the radionuclides in Table 40, iodine isotopes were not included in the exposure dose assessment. The short term constant, short term weathering half-life, and long term weathering half-life for each surface applied to Eq. (3) are shown in Table 41. In this exercise, it was assumed that all radionuclides behave as particulate radionuclides; the parameters related to the weathering effect are those of caesium particulates.

TABLE 40. INITIAL DEPOSITION FOR SURFACE TYPES (kBq/m^2)

Radionuclides	Roof	Paved Road	Wall	Ground	Trees
Cs-137	231.56	187.05	6.27	245.21	204.35
Cs-134	231.56	187.05	6.27	245.21	204.35
Cs-136	39.37	31.80	1.07	41.69	34.74
Ag-110m	0.65	0.52	0.02	0.69	0.57
Te-129m	254.72	205.75	6.89	269.74	224.78

TABLE 41. PARAMETERS RELATED TO THE WEATHERING EFFECT

Surface Types	Constant (a_1)	Short term weathering half-life (τ_1, day)	Long term weathering half-life (τ_2, day)
Roof	0.5	340	2420
Paved Road	0.6	80	10 100
Wall	0.2	365	6935
Ground	0.63	317.6	15 600
Trees	0.8	36.5	36 500

TABLE 42. STANDARD KERMA VALUES FOR A ONE STORY WOODEN HOUSE (pGy/gamma per mm^2)

Energy (MeV)	0.3		0.662		3	
Location and Surface	1st floor	Outside	1st floor	Outside	1st floor	Outside
Window	9.4	7.7	20	16	65	47
Wall	59	57	134	118	450	365
Roof	34	8.3	79	20	270	89
Ground soil	120	259	270	560	1040	1690
Nearby buildings	8	15	17	30	50	82
Trees	29	48	66	100	225	325

TABLE 43. STANDARD KERMA VALUES FOR A FIVE STORY CONCRETE BUILDING (pGy/gamma per mm^2)

Energy	0.3 MeV					0.662 MeV					3 MeV				
Location and surface	1st floor	3rd floor	5th floor	Road	Yard	1st floor	3rd floor	5th floor	Road	Yard	1st floor	3rd floor	5th floor	Road	Yard
Window	2.9	2.9	2.8	8.1	6.6	6.5	6.5	6.5	16	14	24	25	24	48	38
Wall	0.5	0.6	0.4	57	45	2.1	2.1	2	115	89	26	28	26	315	250
Roof	0.0001	0.0001	0.6	0.2	0.3	0.0001	0.01	3.8	0.3	0.4	0.01	0.5	56	0.7	0.8
Courtyard	1.8	0.5	0.25	0.25	252	5.1	1	0.5	0.3	530	39	7.8	3.3	1	1580
Trees	0.33	0.06	0.03	0.01	10	0.9	0.15	0.06	0.02	21	4.7	1.3	0.4	0.08	61
Road	1	0.2	0.09	200	0.2	2.6	0.4	0.15	430	0.4	24	3.1	1.1	1260	4
Wall	2	1.8	1.1	130	57	5.4	5.3	3	270	110	32	33	21	810	320
Roof	0.01	0.05	0.3	2	3	0.08	0.15	0.6	3	4	1	1.5	3.3	8.5	9.5

FIG. 146. The method of changing the kerma value according to the geometry change. (Left, change in geometry of nearby buildings; right, change in geometry from one story to two story building.)

Table 42 shows standard kerma values at two locations (1st floor and outside) given for 6 surfaces for 3 gamma sources with energies of 0.3 MeV, 0.6 MeV and 3 MeV. For the radionuclides which have different gamma energy from these 3 energies, the kerma value was obtained by interpolating the kerma values in Table 42. Additionally, kerma values were modified for considerations of different geometry from the standard. For instance, kerma values were reduced for the nearby buildings by a quarter following the geometry change as shown in the left part of Fig. 146. In the case of a change in the number of floors of a building, the receptor's location were divided into two cases, first floor and second floor (right part of Fig. 146). Also, the kerma values for roof and ground were modified considering the distance between the receptor and the roof and ground. In the case of the workplace, because no specific geometry was provided in the exercise, it was assumed that the workplace is equivalent to the case of 'a large public building' in METRO-K, which consists of a fifth floor commercial or office concrete building facing the other buildings on the other side of the street. The standard kerma values for the workplace are shown in Table 43.

A.10.2.3. Input information

By multiplying the time specific activity concentrations of the radionuclides on each surface by the kerma corresponding to each surface and radionuclide, the exposure dose rate to individuals at a specific time can be evaluated. With the obtained kerma values and radionuclide activity concentrations, the external dose for a person on 31 May 2012, during a day staying at home and for a person at the workplace for a day on 31 May 2012, were evaluated, as shown in Table 44.

As shown in Table 45, the habit data of population groups living in Fukushima were provided in the exercise [14] where a recommended distribution of occupancy values were given. Based on this information, annual doses of Fukushima residents were probabilistically derived. Before applying a distribution for time spent in a location, derived kerma values were applied to habit data with two major premises. To begin with, the lifestyle of the residents follows a normal distribution for the work time. On top of that, the activity time in each space – home and workplace – are independent. Since it seems unreasonable to apply an independent assumption for distributions of time in both home and workplace, the concept of other time (Other time = 24 – (Home + Work)) was introduced for calibrating the time spent at home and at the workplace. If the habit data, especially time spent in each place, has followed a normal distribution, it follows Eqs (4) and (5). That is, normal distributions can be added to each other, and the sum of distributions also follows a normal distribution. In this way, the distribution of daily external exposure doses was calculated for a year starting 1 February 2012 to 31 January

2013 and by adding them together, the dose received for 1 year and the distribution of that dose can both be obtained.

$$Y = \alpha_1 X_1 + \cdots + \alpha_n X_n + \beta \sim N(\mu, \sigma^2) \quad (4)$$

$$\mu = \alpha_1 \mu_1 + \cdots + \alpha_n \mu_n + \beta,\ \sigma^2 = \alpha_1 \sigma_1^2 + \cdots + \alpha_n \sigma_n^2 \quad (5)$$

where:

$$X_i \sim N(\mu_i, \sigma^2),\ 1 \le i \le n\text{: Independent Normal Random Variables}$$

The external doses for the representative persons of the two population groups, indoor workers and outdoor workers, were evaluated with both deterministic and probabilistic approaches. In this study, the annual dose of a representative individual was determined as the 95th percentile of the dose distribution. On the other hand, with the deterministic method, the annual external dose for the representative person was obtained by using the 95th percentile of time spent at all space types in order to conservatively evaluate the dose.

TABLE 44. EXPOSURE DOSE RATE FOR HOME AND WORKPLACE ON 31 MAY 2012 (Sv/day)

Radionuclides	Home		Workplace			
	Indoor	Outdoor	1st story	3rd story	5th story	Outdoor
Cs-137	1.4×10^{-6}	2.52×10^{-6}	4.49×10^{-8}	1.23×10^{-8}	3.93×10^{-8}	1.81×10^{-6}
Cs-134	3.62×10^{-6}	6.44×10^{-6}	1.08×10^{-7}	2.89×10^{-8}	9.92×10^{-8}	4.05×10^{-6}
Cs-136	5.73×10^{-14}	1.0×10^{-13}	1.81×10^{-15}	4.44×10^{-16}	2.04×10^{-15}	5.97×10^{-14}
Ag-110m	9.06×10^{-9}	1.57×10^{-8}	3.04×10^{-10}	8.16×10^{-11}	2.98×10^{-10}	9.79×10^{-9}
Te-129m	1.59×10^{-10}	2.83×10^{-10}	4.51×10^{-12}	1.20×10^{-12}	4.12×10^{-12}	1.72×10^{-10}

TABLE 45. STATISTICS OF TIME SPENT VARIOUS PLACES

Population group (recommended distribution for time spent outdoors from occupancy survey [14])		Total (h)		Time spent per day (h)					
				Home		Workplace		Other	
		Indoor	Outdoor	Indoor	Outdoor	Indoor	Outdoor	Indoor	Outdoor
Indoor worker (Log Normal) N = 11	AM	23.3	0.7	16.1	0.3	6.1	0.0	1.1	0.4
	SD	0.47	0.48	1.45	0.38	0.64	0.06	0.9	0.39
	GM	23.3	0.5	16.1	0.3	6.1	0.1	0.9	0.3
	GSD	1.0	2.6	1.1	1.0	1.1	1.0	2.6	2.5
	95%tile	23.9	1.3	18.0	0.9	6.9	0.1	2.4	1.0
Outdoor worker (Normal) N = 33	AM	16.7	7.3	14.7	2.6	1.1	4.0	0.9	0.8
	SD	2.7	2.7	2.6	2.8	1.7	2.9	0.7	0.8
	GM	15.9	7.4	13.9	1.3	1.3	4.4	0.6	0.5
	GSD	1.2	1.4	1.2	3.4	3.4	1.7	2.7	2.6
	95%tile	20.6	12.1	19.1	8.3	4.5	8.6	2.1	2.6

A.10.3. Results

Figure 147 shows the mean daily doses for indoor and outdoor workers for 300 days from May 2012. Figure 147 indicates that the outdoor workers' doses are 1.5 times higher than those of the indoor workers. This is because the time spent by outdoor workers in buildings at the workplace, which serve a shielding role, is shorter than that of indoor workers. Furthermore, there is a larger standard deviation in the dose received by outdoor workers; this reflects the larger variation in the time spent in the outdoor space compared to the indoor worker.

Table 46 shows the annual external dose assessment results, and Fig. 148 shows the comparative results between the probabilistic and deterministic evaluation of the annual dose obtained from 1 February 2012 to 21 January 2013. The blue and green curves are the cumulative distribution functions (CDF) of the annual dose evaluated by the probabilistic method for the indoor and the outdoor worker, respectively. The dotted vertical lines indicate the point where the CDF curves and the 95th percentile line intersect. The solid vertical lines are the annual dose for the representative persons evaluated by the deterministic method. As can be seen from the graph, the value obtained by the deterministic method was higher than that obtained by the probabilistic method, which means that the deterministic method is more conservative.

In addition, Fig. 149 compares the results of the probabilistic evaluation with actual measurement values. The points on the graph are the measurement results, and our evaluation result shows it is more conservative (i.e. higher dose values) for the population group who received lower doses. However, the gap between the evaluation results and the measurements are within an acceptable level of about 0.5 mSv or less.

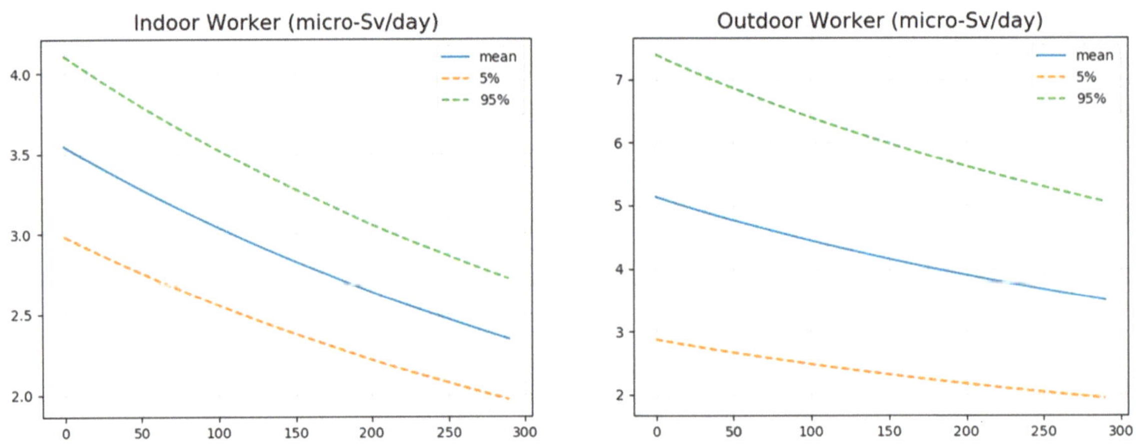

FIG. 147. Daily exposure dose (µSv/day) for indoor and outdoor workers (µSv/day) as a function of days from May 2012.

TABLE 46. ANNUAL EXTERNAL DOSE ASSESSMENT RESULTS

Work type	**Indoor worker (mSv/year)**			**Outdoor worker (mSv/year)**		
Method	Probabilistic		Deterministic	Probabilistic		Deterministic
Annual Dose	Mean		Representative Individual	Mean		Representative Individual
	1.22	1.39	1.50	1.77	2.46	3.37

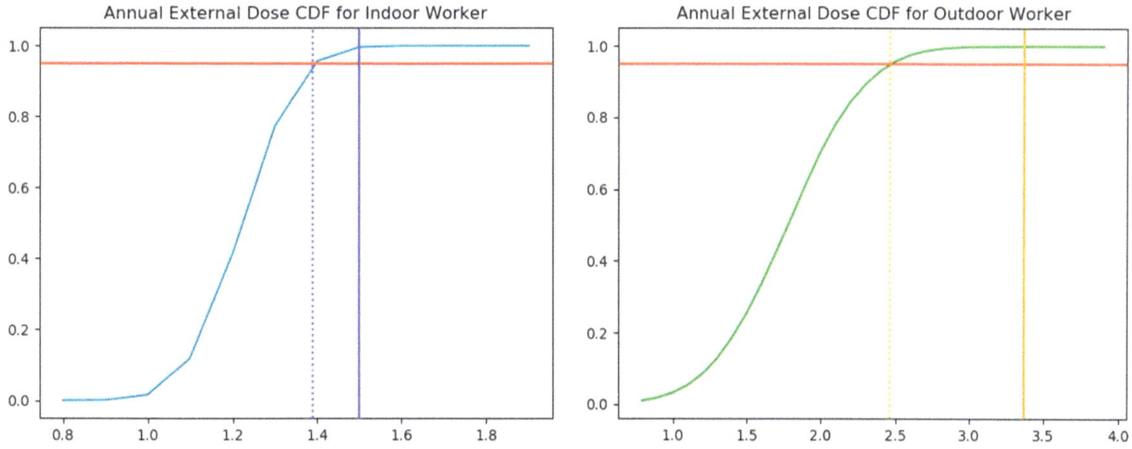

FIG. 148. Comparison of probabilistic and deterministic results. The blue and green curves are the cumulative distribution functions (CDF) of the annual dose in mSv (y-axis) evaluated by the probabilistic method. The dotted vertical lines indicate the point where the CDF curves and the 95th percentile line intersect. The solid vertical lines are the annual dose for the representative persons evaluated by the deterministic method.

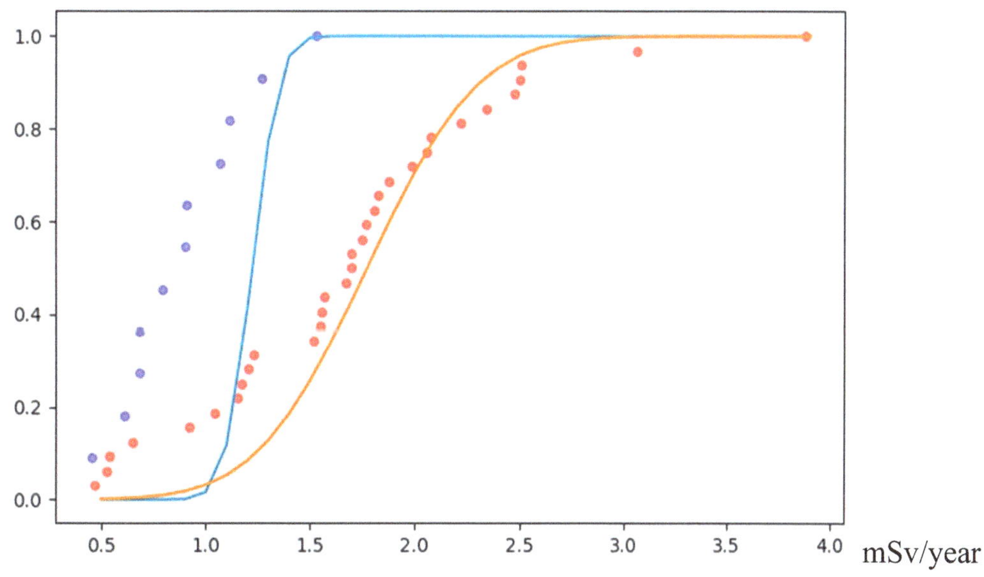

FIG. 149. Comparison of probabilistic assessment results (solid curves) and measurements (circles). The blue curve and circles represent the indoor workers, and the orange curve and circles represent the outdoor workers.

A.10.4. Discussion

In order to conduct the dose assessment for the Fukushima exercise, METRO-K was used and the distribution of annual effective external dose due to deposited radionuclides for indoor workers and outdoor workers were obtained. Furthermore, probabilistic distributions and deterministic results and the measurement values were compared. The daily dose is reduced over time due to decreasing radioactivity caused by radioactive decay and weathering. The dose for a representative person evaluated with the deterministic method is slightly higher than the probabilistic result. Furthermore, the comparative study between the probabilistic results and the measurements showed that the model produced a conservative result. Particularly, the number of people who were exposed to very low doses was expected to be smaller. It is thought that this is due to the assumption of people living in a wooden house, as given in the scenario of the exercise. In Fukushima city, some people live in concrete buildings, which have more shielding than wooden buildings. Therefore, the people who lived in concrete buildings received a lower dose compared to those living in wooden houses.

A.10.5. Acknowledgement

This work was supported by the Nuclear R&D programmes of the Ministry of Science and ICT of Korea. (Grant No. 2017M2A8A4015251).

REFERENCES

[1] INTERNATIONAL ATOMIC ENERGY AGENCY, Environmental Modelling of Remediation of Urban Contaminated Areas, Report of the Urban Remediation Working Group of the EMRAS (Environmental Modelling for Radiation Safety) programme (2012), In: INTERNATIONAL ATOMIC ENERGY AGENCY, Environmental Modelling for Radiation Safety (EMRAS) — A Summary Report of the Results of the EMRAS programme (2003–2007), IAEA-TECDOC-1678, IAEA, Vienna (2012).

[2] INTERNATIONAL ATOMIC ENERGY AGENCY, Assessment of Radioactive Contamination in Urban Areas, IAEA-TECDOC-1941, IAEA, Vienna (2021).

[3] THIESSEN, K.M., et al., Improvement of modelling capabilities for assessing urban contamination: The EMRAS Urban Remediation Working Group, Appl. Radiat. Isot. **66** (2008) 1741–1744. https://doi.org/10.1016/j.apradiso.2007.11.023

[4] THIESSEN, K.M., et al., Modelling of a large-scale urban contamination situation and remediation alternatives, J. Environ. Radioact. **100** (2009) 413–421. https://doi.org/10.1016/j.jenvrad.2009.02.001

[5] THIESSEN, K.M., et al., Modelling the long-term consequences of a hypothetical dispersal of radioactivity in an urban area including remediation alternatives, J. Environ. Radioact. **100** (2009) 445–455. https://doi.org/10.1016/j.jenvrad.2009.02.003

[6] THIESSEN, K.M., ANDERSSON, K.G., CHARNOCK, T.W., GALLAY, F., Modelling remediation options for urban contamination situations, J. Environ. Radioact. **100** (2009) 564–573. https://doi.org/10.1016/j.jenvrad.2009.03.021

[7] THIESSEN, K.M., et al., Assessing emergency situations and their aftermath in urban areas: The EMRAS II Urban Areas Working Group, Radioprotection **46** (2011) S601–S607. https://doi.org/10.1051/radiopro/20116528s

[8] THIESSEN, K.M., et al., Modelling the effectiveness of remediation efforts in contaminated urban areas: An EMRAS II Urban Areas Working Group exercise. In: Proceedings of the WM2015 Conference, 15–19 March 2015, paper #15631, Phoenix, AZ (2015) 1019–1033.

[9] THIESSEN, K.M., et al., Modelling atmospheric dispersion of radiotracers in small-scale, controlled detonations: Validation of dispersion models using field test data, J. Rad. Prot. **42** (2022) 020516. https://doi.org/10.1088/1361-6498/ac66a2

[10] PERIÁÑEZ, R., et al., Mid-range atmospheric dispersion modelling: Intercomparison of simple models in EMRAS-2 project, J. Environ. Radioact. **162–163** (2016) 225–234. https://doi.org/10.1016/j.jenvrad.2016.05.027

[11] INTERNATIONAL ATOMIC ENERGY AGENCY, Assessment of Radioactive Contamination and Effectiveness of Remedial Measures in Urban Environments, IAEA-TECDOC-2001, IAEA, Vienna (2022).

[12] THIESSEN, K.M., et al., Urban Working Groups in the IAEA's Model Testing Programmes: Overview from the MODARIA I and MODARIA II Programmes, J. Rad. Prot. **42** (2022) 020502. https://doi.org/10.1088/1361-6498/ac5173

[13] BOŽNAR, M.Z., et al., Modelling air pollution around nuclear power plants: Validation of dispersion models using tracer data, J. Radiat. Prot. **42** (2022) 020519. https://doi.org/10.1088/1361-6498/ac7a6f

[14] TAKAHARA, S., et al., Assessment of doses in contaminated urban areas: Modelling exercise based on Fukushima data, J. Rad. Prot. **42** (2022) 020517. https://doi.org/10.1088/1361-6498/ac7088

[15] GRAŠIČ, B., PATRYL, L., MLAKAR, P., BOŽNAR, M.Z., KOCIJAN, J., Quality of Weather Forecast for Modelling Air Pollution Dispersion for Nuclear Emergency, HARMO 19: Proceedings, 19th International Conference on Harmonisation within Atmospheric Dispersion Modelling for Regulatory Purposes, 3–6 June 2019, Bruges, Belgium (2019).

[16] PROUZA, Z., et al., Field tests using radioactive matter, Radiat. Prot. Dosim. **139** (2010) 519–531. https://doi.org/10.1093/rpd/ncp299

[17] HOMANN, S.G., HotSpot Health Physics Codes Version 2.07 User's Guide, Lawrence Livermore National Laboratory, LLNL-TM-411345 (2009).

[18] LEE, S., WOLBERG, G., SHIN, S.Y., Scattered data interpolation with multilevel B-splines, IEEE Transactions on Visualization and Computer Graphics **3** (1997) 228–244. https://doi.org/10.1109/2945.620490

[19] ELISEI, G., et al., Experimental campaign for the environmental impact evaluation of Sostanj thermal power plant, Progress Report, ENEL S.p.A, CRAM-Servizio Ambiente, Milano, Italy, C.I.S.E. Tecnologie Innovative S.p.A, Milano, Italy, Institute Jozef Stefan, Ljubljana, Slovenia (1991).

[20] HANFLAND, R., PATTANTYÚS-ÁBRAHÁM, M., RICHTER, C., BRUNNER, D., VOIGT, C., The Lagrangian Atmospheric Radionuclide Transport Model (ARTM) — development, description and sensitivity analysis, Air Quality, Atmosphere and Health (2022). https://doi.org/10.1007/s11869-022-01188-x

[21] MLAKAR, P., et al., "Šostanj" Data Set for Validation of Models over Very Complex Terrain. HARMO 16: Proceedings, 16th International Conference on Harmonisation within Atmospheric Dispersion Modelling for Regulatory Purposes, 8–11 September 2014, Varna, Bulgaria (2014).

[22] SAHA, S., et al., The NCEP Climate Forecast System Reanalysis, Bull. Am. Meteorol. Soc. **91** (2010) 1015–1057.

[23] SAHA S., et al., The NCEP Climate Forecast System Version 2, J. Clim. **27** (2014) 2185–2208.

[24] BADESCU, V., et al., Accuracy and sensitivity analysis for 54 models of computing hourly diffuse solar irradiation on clear sky, Theor. Appl. Climatol. **111** (2013) 379–399. https://doi.org/10.1007/s00704-012-0667-1

[25] KOCIJAN, J., GRADIŠAR, D., BOŽNAR, M.Z., GRAŠIČ, B., MLAKAR, P., On-line algorithm for ground-level ozone prediction with a mobile station, Atmos. Environ. **131** (2016) 326–333. https://doi.org/10.1016/j.atmosenv.2016.02.012

[26] BOŽNAR, M.Z., GRAŠIČ, B., MLAKAR, P., SOARES, J.R., DE OLIVEIRA, A.P., COSTA, T.S. Radial frequency diagram (sunflower) for the analysis of diurnal cycle parameters: solar energy application, Appl. Energy **154** (2015) 592–602. https://doi.org/10.1016/j.apenergy.2015.05.055

[27] BOŽNAR, M.Z., GRAŠIČ, B., MLAKAR, P., GRADIŠAR, D., KOCIJAN, J., The use of a new diagram for the analysis of the daily cycles in the air-pollution data, Int. J. Environ. Pollut. **62** (2017) 385–394. https://doi.org/10.1504/ijep.2017.089422

[28] DEE, D.P., et al., The ERA-Interim reanalysis: configuration and performance of the data assimilation system, Q. J. R. Meteorol. Soc. **137** (2011) 553–597. https://doi.org/10.1002/qj.828

[29] KOVALETS, I.V., et al., Influence of the diagnostic wind field model on the results of calculation of the microscale atmospheric dispersion in moderately complex terrain, Atmos. Environ. **79** (2013) 29–35. https://doi.org/10.1016/j.atmosenv.2013.06.015

[30] KOROLEVYCH, V.Y., KIM, S.B., Relation between the tritium in continuous atmospheric release and the tritium contents of fruits and tubers, J. Environ. Radioact. **118** (2013) 113–120. https://doi.org/10.1016/j.jenvrad.2012.12.004

[31] UNITED STATES NUCLEAR REGULATORY COMMISSION, Atmospheric Dispersion Models for Potential Accident Consequence Assessments at Nuclear Power Plants, USNRC Regulatory Guide 1.145 (1979).

[32] TINARELLI, G., ANFOSSI, D., BIDER, M., FERRERO, E., TRINI CASTELI, S., A new high performance version of Lagrangian particle dispersion model SPRAY, some case studies, Air pollution modelling and its Applications XIII, S.E. Gryning and E. Batchvarova (Eds), Kluwer Academic / Plenum Press, New York (2000) 499–507. https://doi.org/10.1007/978-1-4615-4153-0_51

[33] TAKAHARA, S., IIJIMA, M., SHIMADA, K., KUSHIDA, T., SHIRATORI, Y., Development of Deterministic Approach to Assess Doses to the Public from External Exposures in the Areas Contaminated by the Fukushima Daiichi Nuclear Power Station Accident, Japan Atomic Energy Agency, Ibaraki, JAEA-Research 2014-024 (2014).

[34] TAKAHARA, S., IIJIMA, M., WATANABE, M., Assessment model of radiation doses from external exposure to the public after the Fukushima Daiichi nuclear power plant accident, Health Phys. **118** (2020) 664–677. https://doi.org/10.1097/hp.0000000000001176

[35] EUROPEAN COMMISSION, FOOD AND AGRICULTURE ORGANIZATION OF THE UNITED NATIONS, INTERNATIONAL ATOMIC ENERGY AGENCY, INTERNATIONAL LABOUR ORGANIZATION, OECD NUCLEAR ENERGY AGENCY, PAN AMERICAN HEALTH ORGANIZATION, UNITED NATIONS ENVIRONMENT PROGRAMME, WORLD HEALTH ORGANIZATION, Radiation Protection and Safety of Radiation Sources: International Basic Safety Standards, IAEA Safety Standards Series No. GSR Part 3, IAEA, Vienna (2014), https://doi.org/10.61092/iaea.u2pu-60vm

[36] FUKUSHIMA PREFECTURE, Reading of air dose rate for the monitoring post in Momijiyama (Fukushima City) (2011). Available online only at: https://emdb.jaea.go.jp/emdb_old/en/portals/1010401000/

[37] UNITED NATIONS SCIENTIFIC COMMITTEE ON THE EFFECTS OF ATOMIC RADIATION, Sources and Effects of Ionizing Radiation, UNSCEAR 1993 Report to the General Assembly, with Scientific Annexes, United Nations, New York (1993). https://doi.org/10.18356/2c4203dc-en

[38] FUKUSHIMA PREFECTURE, Result of dose rate measurement in March 2011 (2011) (Data collected from monitoring post of Fukushima Prefecture, online only (in Japanese) Available at: http://www.atom-moc.pref.fukushima.jp/old/monitoring/monitoring201103/201103_mpdata.html

[39] JAPAN METEOROLOGICAL AGENCY, Download of meteorological data of the past, JMA (2011), online only (in Japanese). Available at: http://www.data.jma.go.jp/gmd/risk/obsdl/index.php

[40] UNITED NATIONS SCIENTIFIC COMMITTEE ON THE EFFECTS OF ATOMIC RADIATION, UNSCEAR 2013 Report, Volume 1, Report to the General Assembly, Scientific Annex A: Levels and Effects of Radiation Exposure Due to the Nuclear Accident after the 2011 Great East-Japan Earthquake and Tsunami, UNSCEAR, United Nations, New York (2014). https://doi.org/10.18356/78a5ff52-en

[41] YOSHIMURA, K., SAITO, K., FUJIWARA, K., Distribution of ^{137}Cs on components in urban area four years after the Fukushima Dai-ichi nuclear power plant accident, J. Environ. Radioact. **178–179** (2017) 48–54. https://doi.org/10.1016/j.jenvrad.2017.07.021

[42] FURUTA, T., TAKAHASHI, F., Analyses of Radiation Shielding and Dose Reduction in Buildings for Gamma-rays Emitted from Radioactive Cesium in Environment Discharged by a Nuclear Accident, Japan Atomic Energy Agency, Ibaraki, JAEA-Research 2014-003 (2014).

[43] YOSHIDA-OHUCHI, H., MATSUDA, N., SAITO, K., Review of reduction factors by buildings for gamma radiation from radiocaesium deposited on the ground due to fallout, J. Environ. Radioact. **187** (2018) 32–39. https://doi.org/10.1016/j.jenvrad.2018.02.006

[44] CHARNOCK, T.W., LANDMAN, C., TRYBUSHNYI, D., IEVDIN, I., European model for inhabited areas – ERMIN 2, Radioprotection **51(HSI)** (2016) S23-S25. https://doi.org/10.1051/radiopro/2016006

[45] YU, C., et al., User's Manual for RESRAD-BUILD Version 3, Argonne National Laboratory, Argonne, IL, ANL/EAD/03-1 (2003).

[46] ANDERSSON, K.G., et al., Airborne contamination in the indoor environment and its implications for dose. Risø National Laboratory, Roskilde, Denmark, Risø-R-1462(EN) (2004).

[47] YU, C., CHENG, J.-J., KAMBOJ, S., DOMOTOR, S., WALLO, A., Preliminary Report on Operational Guidelines Developed for Use in Emergency Preparedness and Response to a Radiological Dispersal Device Incident, Argonne National Laboratory, Argonne, IL, ANL/EVS/TM/09-1, DOE/HS-0001, Interim Final (2009). https://doi.org/10.2172/925331

[48] KINASE, S., TAKAHASHI, T., SAITO, K., Long-term predictions of ambient dose equivalent rates after the Fukushima Daiichi nuclear power plant accident, J. Nucl. Sci. Technol. **54** (2017) 1345–1354. https://doi.org/10.1080/00223131.2017.1365659

[49] BUNZL, K., SCHIMMACK, W., ZELLES, L., ALBERS, B.P., Spatial variability of the vertical migration of fallout ^{137}Cs in the soil of a pasture, and consequences for long-term predictions, Radiat. Environ. Biophys. **39** (2000) 197–207. https://doi.org/10.1007/s004110000062

[50] ANDERSSON, K.G., et al., Physical Countermeasures to Sustain Acceptable Living and Working Conditions in Radioactively Contaminated Residential Areas, Risø National Laboratory, Roskilde, Denmark, Risø-R-1396(EN) (2003).

[51] GALE, H.J., HUMPHREYS, D.L., FISHER, E.M., Weathering of Caesium-137 in soil, Nature **201** (1964) 257–261. https://doi.org/10.1038/201257a0

[52] CHARNOCK, T., Enhancement of the ERMIN urban dose and remediation model to account for physicochemical properties of contamination, Radioprotection **51(HS2)** (2016) S105–S107. https://doi.org/10.1051/radiopro/2016042

[53] PETERSON, S.-R., CHOUHAN, S., HEINMILLER, B., KOCH, J., CHERURB-95: Urban Contamination and Dose Model, Research report prepared for the Atomic Energy Control Board, Canada (1995).

[54] SAITO, K., ISHIGURE, N., PETOUSSI-HENNSS, N., SCHLATTL, H., Effective dose conversion coefficients for radionuclides exponentially distributed in the ground, Radiat. Environ. Biophys. **51** (2012) 411–423. https://doi.org/10.1007/s00411-012-0432-y

[55] JONES, J.A., Investigation of the effects of building spacings and configurations of trees for dose rates in the CONDO v3.1 project, Chilton, UK, NRPB-EA/11/2004 (2004).

[56] MECKBACH, R., JACOB, P., PARETZKE, H.G., Gamma exposures due to radionuclides deposited in urban environments, Part I: Kerma rates from contaminated urban surfaces, Radiat. Prot. Dosim. **25** (1988) 167–179. https://doi.org/10.1093/oxfordjournals.rpd.a080369

[57] ECKERMAN, K.F., RYMAN, J.C., External Exposure to Radionuclides in Air, Water, and Soil, Exposure to Dose Coefficients for General Application, Based on the 1987 Federal Radiation Protection Guidance, U.S. Environmental Protection Agency, Office of Radiation and Indoor Air, Washington, D.C., EPA 402-R-93-076, Federal Guidance Report No. 12 (1993).

[58] GOLIKOV, V.Y., BALONOV, M.I., JACOB, P., External exposure of the population living in areas of Russia contaminated due to the Chernobyl accident, Radiat. Environ. Biophys. 41 (2002) 185–193. https://doi.org/10.1007/s00411-002-0167-2

[59] INTERNATIONAL COMMISSION ON RADIOLOGICAL PROTECTION, Conversion Coefficients for Use in Radiological Protection against External Radiation, ICRP Publication 74, Ann. ICRP 26 (3–4) (1996).

[60] NISBET, A., et al., Generic handbook for assisting in the management of contaminated inhabited areas in Europe following a radiological emergency, EURANOS(CAT1)-TN(09)-03 (2010).

[61] INTERNATIONAL COMMISSION ON RADIOLOGICAL PROTECTION, Assessing Dose of the Representative Person for the Purpose of Radiation Protection of the Public, ICRP Publication 101, Ann. ICRP 36 (3) (2006). https://doi.org/10.1016/j.icrp.2006.09.003

[62] FINLEY, B., PAUSTENBACH, D., The benefit of probabilistic exposure assessment: Three case studies involving contaminated air, water, and soil, Risk Anal. **14** (1994) 53–73. https://doi.org/10.1111/j.1539-6924.1994.tb00028.x

[63] HOMANN, S.G., ALUZZI, F., HotSpot Health Physics Codes Version 3.0 User's Guide. Lawrence Livermore National Laboratory, LLNL-SM-636474 (2014).

[64] WALTER, H., HEINRICH, G., Effects from urban structures to atmospheric dispersion models in decision support systems for nuclear emergencies, 17th International Conference on Harmonisation within Atmospheric Dispersion Modelling for Regulatory Purposes, 9–12 May 2016, Budapest, Hungary (2016).

[65] VEREIN DEUTSCHER INGENIEURE, Dispersion of Pollutants in the Atmosphere: Determination of the Plume Rise, Technical Rule VDI 3782 Part 3, Beuth (1985).

[66] RICHTER, C., SOGALLA, M., THIELEN, H., MARTENS, R., ARTM Atmospheric Radionuclide Transport Model Version 2.8.0 Modellbeschreibung, In: Ergänzende Untersuchungen zur Validierung des Atmosphärischen Radionuklid-Transport-Modells (ARTM). GRS-394, Gesellschaft für Anlagen- und Reaktorsicherheit (2015) (in German).

[67] RASKOB, W., TRYBUSHNYI, D., IEVDIN, I., ZHELEZNYAK, M., JRODOS: Platform for improved long term countermeasures modelling and management, Radioprotection **46** (2012) S731–S736. https://doi.org/10.1051/radiopro/20116865s

[68] RASKOB, W., LANDMAN, C., TRYBUSHNYI, D., Functions of decision support systems (JRodos as an example): overview and new features and products, Radioprotection **51** (2016) S9–S11. https://doi.org/10.1051/radiopro/2016015

[69] JANICKE CONSULTING, Dispersion Model LASAT – Reference Book for Version 3.4 LASAT Reference Book 3.4I (2020).

[70] THYKIER-NIELSEN, S., DEME, S., MIKKELSEN, T., Description of the Atmospheric Dispersion Module RIMPUFF, Roskilde, Risø National Laboratory, RODOS(WG2)-TN(98)-02 (1999).

[71] PETOUSSI-HENSS, N., SCHLATTL, H., ZANKL, M., ENDO, A., SAITO, K., Organ doses from environmental exposures calculated using voxel phantoms of adults and children, Phys. Med. Biol. **57** (2012) 5679–5713. https://doi.org/10.1088/0031-9155/57/18/5679

[72] CHARNOCK, T.W., The European model for inhabited areas (ERMIN) – developing a description of the urban environment, Radioprotection **45**(5) (2010) S55-S61. https://doi.org/10.1051/radiopro/2010016

[73] CHARNOCK, T.W., Addressing the uncertainties in urban/inhabited scenarios, Appendix 2, ERMIN Uncertainty, EJC CONCERT EURATOM H2020 – 662287, D9.20 (2018).

[74] JONES, J.A., et al., Description of the Modelling of Transfer and Dose Calculations within ERMIN v1.0 and Associated Data Libraries, Chilton, UK, EURANOS(CAT2)-TN(05)-04 (2007).

[75] ANDERSSON, K.G., Addressing the uncertainties in urban/inhabited scenarios, Appendix 1, Urban scenario parameter uncertainty, EJC CONCERT EURATOM H2020 – 662287, D9.20 (2018).

[76] HWANG, W.T., KIM, E.H., JEONG, H.J., SUH, K.S., HAN, M.H., A model for radiological dose assessment in an urban environment, J. Radiat. Prot. **32** (2007) 1–8.

CONTRIBUTORS TO DRAFTING AND REVIEW

Božnar, M.Z.	MEIS d.o.o., Republic of Slovenia
Brown, J.	International Atomic Energy Agency
Charnock, T.W.	UK Health Security Agency, United Kingdom
Federspiel, L.	National Emergency Operations Centre, Switzerland
Grašič, B.	MEIS d.o.o., Republic of Slovenia
Halsall, C.	International Atomic Energy Agency
Heinrich, G.	Bundesamt für Strahlenschutz, Germany
Helebrant, J.	National Radiation Protection Institute, Czech Republic
Kamboj, S.	Argonne National Laboratory, United States of America
Kocijan, J.	Jožef Stefan Institute, Republic of Slovenia
Korolevych, V.	Canadian Nuclear Laboratories, Canada
Kuča, P.	National Radiation Protection Institute, Czech Republic
Lee, J.	Korea Atomic Energy Research Institute, Republic of Korea
Mancini, F.	SOGIN S.p.A., Italy
Mlakar, P.	MEIS d.o.o., Republic of Slovenia
Patryl, L.	Commissariat à l'Energie Atomique, France
Pattantyús-Ábrahám, M.	Bundesamt für Strahlenschutz, Germany
Popović, D.	MEIS d.o.o., Slovenia
Silva, K.	Thailand Institute of Nuclear Technology, Thailand
Takahara, S.	Japan Atomic Energy Agency, Japan
Thiessen, K.M.	Oak Ridge Center for Risk Analysis, United States of America
Walter, H.	Bundesamt für Strahlenschutz, Germany
Yankovich, T.	International Atomic Energy Agency
Yu, C.	Argonne National Laboratory, United States of America

LIST OF PARTICIPANTS

AbuHejleh, H.	Jordanian Atomic Energy Commission, Jordan
Alotaibi, K.	Environment Protection and Meteorology General Authority, Saudi Arabia
Amer, H.	Egyptian Nuclear and Radiological Regulatory Authority, Egypt
Božnar, M.Z	Modelling, Meteorology, Measurement, Environmental Information Systems Engineering, Republic of Slovenia
Burbidge, C.	Environmental Protection Agency, Republic of Ireland
Charnock, T.W.	UK Health Security Agency, United Kingdom
Cheng, W.	China Institute of Atomic Energy, People's Republic of China
Encian, I.	National Commission for Nuclear Activities Control, Romania
Farook, A.	Pakistan Institute of Engineering and Applied Sciences, Pakistan Atomic Energy Agency, Pakistan
Federspiel, L.	National Emergency Operations Centre, Switzerland
Grašič, B.	Modelling, Meteorology, Measurement, Environmental Information Systems Engineering, Republic of Slovenia
Gu, Z.	China Institute for Radiation Protection, People's Republic of China
Halleröd, J.	Vattenfall AB, Sweden
Heidary, P.	Iran Nuclear Regulatory Authority, Atomic Energy Organization of Iran, Islamic Republic of Iran
Heinrich, G.	Bundesamt für Strahlenschutz, Germany
Helebrant, J.	National Radiation Protection Institute, Czech Republic
Huutoniemi, T.	Vattenfall AB, Sweden
Hwang, W.T.	Korea Atomic Energy Research Institute, Republic of Korea
Irfan, N.	Pakistan Institute of Engineering and Applied Sciences, Pakistan Atomic Energy Agency, Pakistan
Kamboj, S.	Argonne National Laboratory, United States of America
Kocijan, J.	Jožef Stefan Institute, Republic of Slovenia
Korenová, Z.	Nuclear Regulatory Authority of the Slovak Republic, Slovakia
Korolevych, V.	Canadian Nuclear Laboratories, Canada
Kuča, P.	National Radiation Protection Institute, Czech Republic

Kwamena, N.	Canadian Nuclear Safety Commission, Canada
Lee, J.	Korea Atomic Energy Research Institute, Republic of Korea
Liao, Y.	Ministry of Environmental Protection, People's Republic of China
Little, A.	Defence Academy of the United Kingdom, United Kingdom
Mancini, F.	SOGIN S.p.A., Italy
Mlakar, P.	Modelling, Meteorology, Measurement, Environmental Information Systems Engineering, Republic of Slovenia
Nadeem, Q.	Pakistan Institute of Engineering and Applied Sciences, Pakistan Atomic Energy Agency, Pakistan
Patryl, L.	Commissariat à l'Energie Atomique, France
Pattantyús-Ábrahám, M.	Bundesamt für Strahlenschutz, Germany
Psaltaki, M.	National Technical University of Athens, Greece
Reisin, T.G.	Israeli Atomic Energy Commission, Israel
Sdouz, G.	Consultant, Austria
Shubayr, N.	University of Michigan United States of America
Silva, K.	Thailand Institute of Nuclear Technology, Thailand
Sim, E.	DSO National Laboratories, Singapore
Staudt, C.	Karlsruhe Institute of Technology, Germany
Sun, H.	Nuclear and Radiation Safety Centre, People's Republic of China
Takahara, S.	Japan Atomic Energy Agency, Japan
Tawfik, F.	Egyptian Nuclear and Radiological Regulatory Authority, Egypt
Tay, B.K.	DSO National Laboratories, Singapore
Thiessen, K.M.	Oak Ridge Center for Risk Analysis, United States of America
Thorn, J.	Vattenfall AB, BA Generation, Sweden
Tomas, J.	National Institute of Public Health and the Environment, Netherlands
Walter, H.	Bundesamt für Strahlenschutz, Germany
Wu, Q.	Tsinghua University, People's Republic of China
Yankovich, T.	International Atomic Energy Agency
Yu, C.	Argonne National Laboratory, United States of America

MODARIA II Technical Meetings, IAEA Headquarters, Vienna

31 October – 4 November 2016, 30 October – 3 November 2017, 22–25 October 2018, 21–24 October 2019

Interim Working Group Meetings, MODARIA II Working Group 2

Vienna, Austria, 22–24 May 2017, 23–25 April 2018, 1–3 April 2019

IAEA
International Atomic Energy Agency

CONTACT IAEA PUBLISHING

Feedback on IAEA publications may be given via the on-line form available at:
www.iaea.org/publications/feedback

This form may also be used to report safety issues or environmental queries concerning IAEA publications.

Alternatively, contact IAEA Publishing:

Publishing Section
International Atomic Energy Agency
Vienna International Centre, PO Box 100, 1400 Vienna, Austria
Telephone: +43 1 2600 22529 or 22530
Email: sales.publications@iaea.org
www.iaea.org/publications

Priced and unpriced IAEA publications may be ordered directly from the IAEA.

ORDERING LOCALLY

Priced IAEA publications may be purchased from regional distributors and from major local booksellers.

25-02088E

MIX
Papier | Fördert gute Waldnutzung
FSC® C083411

Zeitfracht Medien GmbH
Ferdinand-Jühlke-Straße 7
99095 Erfurt, Deutschland
produktsicherheit@kolibri360.de